Genealogy of Obedience

Human-Animal Studies

VOLUME 20

The titles published in this series are listed at brill.com/has

Genealogy of Obedience

Reading North American Dog Training Literature, 1850s–2000s

By

Justyna Włodarczyk

BRILL

LEIDEN | BOSTON

Two sections of Chapter 7 (titled in the book: "Garrett, Foucault and Radical Behaviorism" and "Beyond Behaviorism") have been published as: "Be More Dog: The Human-Canine Relationship in Contemporary Dog Training Methodologies," *Performance Research* 22.2 (2017): 40–47. DOI:10.1080/13528165.2017.1315962
One section of chapter 6 (titled in the book: "Disciplining Affects: The Dog Whisperer") has been published as "Disciplining Affects: The Dog Whisperer," *Fictions. Studi sulla narratività*, XVII, (2018): 49–62. DOI:10.19272/201806901004

Cover illustration: Front image, "Sitting Pretty" courtesy of the author. Background image: Ten Lords a' Leaping (1930). Photographer: George Jackman for Queensland Newspapers Pty Ltd. Retrieved by State Library of Queensland, Australia [No restrictions], via Wikimedia Commons.

Library of Congress Cataloging-in-Publication Data

Names: Włodarczyk, Justyna, author.
Title: Genealogy of obedience : reading North American dog training literature, 1850s–2000s / by Justyna Wlodarczyk.
Description: Boston : Brill, [2018] | Series: Human-animal studies, ISSN 1573-4226 ; volume 20 | Includes bibliographical references and index.
Identifiers: LCCN 2018031879| ISBN 9789004380288 (hardback) | ISBN 9789004380295 (ebook)
Subjects: LCSH: Dogs—Training—United States—History—19th century. | Dogs—Training—United States—History—20th century. | Dogs—Training—United States—History—21st century.
Classification: LCC SF431 .W655 2018 | DDC 636.7/0835—dc23
LC record available at http://lccn.loc.gov/2018031879

Typeface for the Latin, Greek, and Cyrillic scripts: "Brill." See and download: brill.com/brill-typeface.

ISSN 1573-4226
ISBN 978-90-04-38028-8 (hardback)
ISBN 978-90-04-38029-5 (e-book)

This book is printed on acid-free paper and produced in a sustainable manner.

Contents

Acknowledgments

This book would not have been possible without various forms of support that I have received over the years of work on the project. Most significantly, I have been the beneficiary of two fellowships that have allowed me to develop my ideas in supportive environments: the Animals and Society/Wesleyan Animal Studies Fellowship in Human-Animal Studies (2015), which resulted in the first drafts of chapters 1 and 2 of this book, and the Fulbright Senior Research Award (2016/2017), which took me to the Forum for International US Studies at the University of Illinois at Urbana-Champaign, where I spent a wonderfully productive year devoted solely to this project. The supportive environment created by IFUSS creators Jane Desmond and Virginia Rodriguez is second to none; I cannot say enough good things about the atmosphere, the office space, university facilities, easy access to archives, continuous and lively intellectual debates and personal support. I have benefitted from the feedback of other IFUSS fellows during regular workshops and from almost daily conversations over coffee with David Schraig, program coordinator for IFUSS. I would also like to thank my home university, The University of Warsaw, for readily granting me academic leave (multiple times) to pursue this project. My Polish colleagues – especially Marek Paryż, Zuzanna Ładyga, Anna Pochmara, Joanna Ziarkowska, Ewa Łuczak – have always been there for me when I needed them. Zuzanna, who has never lived with a dog, probably knows more about dog training than many dog trainers do.

I have received an outpouring of support from animal studies colleagues whom I contacted by e-mail and at conferences, asking for clarification of their own research and for feedback on my ideas. I have benefitted immensely from a mostly online relationship with the late prof. Raymond Coppinger, whose input into chapters 2 and 3 was significant and whose work has been very inspirational for me. Drafts of most of the chapters were presented at various conferences; the one that I remember the most vividly as a truly welcoming space for intellectual discussion being the *Living with Animals* conference at Eastern Kentucky University (2017), organized by Bob Mitchell and Radhika Makecha. Michał Pręgowski's unwavering belief in my competence to write about training as well as his own enthusiasm and energy have helped me maintain my own motivation.

I have been immersed in the dog community for many years and would also like to thank my non-academic friends for the many discussions we have had about training, which have often been amazingly fruitful. It would be hard to list all my "dog friends" whose influence on this book is considerable but

Magda Gmitrzak, Basia Kaczmarczyk, Magda Widlak, Karolina Nowak, Maria Pajzderska are certainly among them. I hope you enjoy this book as well.

Finally, I would like to thank my family for their ongoing support of the endeavor and willingness to relocate to Illinois with me, even if somewhat grudgingly. Irmina and Janusz, I am glad you survived the cornfields and the wind. To my mom, for serving as housesitter, babysitter, petsitter, and number one fan. I cannot say that my dogs ever expressed any kind of enthusiastic support, but they have gracefully survived the turmoil of this period of intense writing and editing, which they probably associate with less attention (and training!) for them. The memory of the relationship I had with my dearest Uma, whom we lost in 2015, keeps me grounded in the belief that humans and canines can truly be companion species and that training together can be a vehicle for increasing interspecies understanding. The presence of the dogs I currently live with – Malcolm, Ida, Migo and Flora – keeps me relatively sane and humble.

Preface

The idea for this book began with something of a conceptual experiment: the application of Foucault to dog training, in an attempt to explain the difference between Foucault's disciplinary society and the post-Foucauldian term "society of control," as employed by Deleuze (1992), Rose (1999a) and Hardt and Negri (2001). To me – someone with a history of involvement with dogs – dog training seemed a perfect example to illustrate the waning of the classic disciplinary regime: we go from leashes and obedience schools to cookies and individually-tailored behavior modification programs; from hands-on to hands-off methods; from six-week-long group classes to perpetual training at home. I was discussing *Discipline and Punish* with a group of undergraduates who were not "dog people." Still, they seemed to find the example useful and contributed stories of their own: it seemed everyone had at least heard of a friend or family member's adventures in the world of dog training. Because the students' response was overwhelmingly positive, I checked if anyone had written about this shift in dog training practices using a Foucauldian vocabulary. No one had. As I was re-reading Foucault's *Technologies of the Self* (1988) with a group of graduate students, the analogy between the self-control practiced in Greek *askesis* and self-control as one of the foundational exercises in reinforcement-based dog training again taunted me with possibilities for further exploration. Then there was Diogenes the Cynic, who literally led a dog's life; an iconic figure for animal studies and for Foucauldian biopolitics. When I discovered that Foucault had read Skinner, the giant of American behaviorism and potentially the greatest influence on dog training today, I was hooked. I knew I had the making of a full-fledged project that was both practical and intellectually productive: it was an interpretive strategy that could explain what might otherwise remain baffling. Reading dog training literature through a Foucauldian framework made sense; it flowed; it explained certain conundrums that would otherwise be left unanswered. Why is an obscure German army colonel, who never set foot in the US, considered the father of modern dog training in North America? Why do contemporary positive dog trainers swear by B.F. Skinner, who trained pigeons to carry out suicide missions? Why were mid-twentieth-century American trainers so opposed to teaching dogs tricks? Why did William Koehler, who is now condemned for excessively cruel methods, receive awards from the American Humane Association and the Dog Writers' Association of America?

Of course, my interest in tracing the genealogy of dog training has a personal dimension, even if the personal voice is absent from my argument in this book. This analysis is fueled by my own involvement in the dog world: I am – and have been ever since I remember – a "dog person." In an attempt to explain how some "pet owners" become "dog people," performance artist Holly Hughes writes: "Our lives revolve around our dogs. We don't just rig our teaching schedules around the need to let them out every few hours. We train them, and the majority of our weekends are built around travelling to soccer arenas and stables to compete in dog shows" (Hughes 2015, 212). This has been the story of my life. My greatest joys and my deepest heartbreaks have been related to dogs. My involvement with dogs has, to a large extent, defined how and where I live my life, how my day is organized, whom I call kin, where and how I travel (I have moved with my dogs across continents, multiple times). It has also shaped my academic path. The existence of the animal turn in the academy has, of course, served as an enabler. An early mentor, in my pre-animal studies days, once asked me: "What would you be researching if you weren't doing this?" I did not hesitate. My answer was immediate: "Dogs." At the time, it did not seem possible, at least to me, to carry out serious research within American Studies in Europe on a topic that had such a trivial ring to – as I felt – everyone but myself. Now, having a community of scholars to fall back on, the animal studies community, makes it much more feasible.

Yet, it seems that dog training is a tricky topic among animal studies scholars. This suspicion towards training is understandable, particularly in the US context. Firstly, outside of a religious framework, obedience seems antithetical to the American experience. Beginning with Emerson and Thoreau, America explicitly values individualism and non-conformity. In fact, the subsequent emancipation of oppressed groups has taken place through acts of civil disobedience. If we conceptualize the emergence of animal studies – even if it does not always directly connect to the discourse of animal rights – as a discourse that probes and questions the status of animals in contemporary society, it is not surprising that it is not going to be interested in an exploration of methods for enforcing obedience. On the contrary, resistance looms as a more potent topic for analysis. Yet obedience, as it relates to the human-canine relationship, has a history: this book is the story of the rise and fall of obedience as a conceptual category in human-canine relations. Today, obedience – understood as unconditional submission to the will of the other: a definition Foucault uses to explain the difference between Cynic *askesis* and Christian ascetisism (Foucault 2008, 320) – is long dead even among those who participate in obedience competitions. One of the goals of this book is to show that the very impulses behind the animal turn have also shaped the changes in contemporary dog training. As a corollary, I also want to show how dog training responds to bigger shifts in conceptualizing the world: how it is not static. And again, Foucault is amazingly useful for making this point. Relations of power are not immutable and unchanging. As academics, we would violently object to describing the contemporary student-teacher relationship in terms of obedience to authority. Hopefully, we wish to see ourselves as fostering a potential for critical thinking and civil disobedience. Yet somehow, when we begin speaking of interspecies relationships, the radical potential of the student-teacher relationship is hardly ever taken up. I hope to show that contemporary dog training tries to be a pedagogy more radical than the one structuring many (human) teacher-student relationships.

Secondly, a lot of effort has been placed (and rightly so!) on revealing the oppressive aspects of entertainment involving wild animals in circuses and marine mammal parks. This has resulted in a significant body of recent historical scholarship on the training of wild animals, e.g. Susan Nance's *Entertaining Elephants* (2013) or Peta Tait's *Fighting Nature* (2016). Somewhat ironically, the training of animals who have been closest to humans has not yet received the same kind of attention.[1] It seems as if the training of dogs has become the dominion of animal behavior and cognition laboratories; the highly technical studies produced by these labs and chapters in textbooks of behavioral science and ethology. These – while their importance cannot be overstated – are preoccupied with which techniques work and not with why people believe/have believed certain techniques work: the question a cultural studies scholar finds much more interesting. While involved in ethical projects of their own, these studies do not offer a space for humanities-based reflection on the complexity of the training relationship. There do exist exceptions, i.e. analyses of dog training coming from the humanities: Donna Haraway's books (2003; 2008) and the entire trend she could be seen as commencing are supremely important for understanding the contemporary training relationship as exceeding the framework of obedience. Haraway is also responsible for opening up the field: for turning dog training into a legitimate subject of humanities-based scholarly analyses, for which I am forever grateful.

In other words, this project has been a way of making sense of my own involvement in an activity that certainly exceeds the definition of a hobby. In fact, as I found myself buried deep in obscure training manuals from the past, I felt my life had gone full circle. I remember myself as a dog-crazy girl, begging my parents for a dog that I could call my own and reading all dog training books I could lay my hands on. As a ten-year-old I read them as unquestioningly as

1 My use of the pronoun who to refer to animals is intentional. I use it consistently throughout the book. I also refer to the dog using the personal pronoun "he"; chosen for the sake of consistency. I tried interchanging "he" and "she" but found it too chaotic.

I would read a school textbook. If experts say this is how it must be done, this is how it must be done. My own development as a dog trainer, after I finally came back from the dog pound with a scruffy looking pup of my own, resembles the narratives recounted by many of the trainers whose work is analyzed here. Shaped by the same belief in the power of love that so many girls (and increasingly more boys) are raised with, I could not just gloss over the fact that the training techniques I was required to use seemed to have little to do with the love I was feeling: I felt guilty, I felt uneasy with what I was doing. Clicker training, rooted in Skinner's radical behaviorism, seemed a lifesaver: offering hope to improve the bond between myself and my dog while providing me with a comprehensive technology for shaping behaviors. Preoccupied with my dogs' emotional well-being, I became involved in dog sports, which offered themselves as an outlet for canine energy in an urbanized world. My dogs and I have been very successful in a number of activities, but I have never been able to embrace the mindset required for true competitiveness: that winning edge, although, I admit, the euphoria co-experienced – I can only hope – with a dog upon the completion of a clean agility run or a winning disc freestyle routine has accounted for some of the most intense experiences of *jouissance* in my life. Still, the academic in me always maintained a level of detachment from the activities I was engaged in. As I allowed myself to turn my research in the direction of animal studies, I felt my history of involvement with dog training provided me with a unique perspective. I do not refer to these experiences explicitly in this book: my methodology is based on a reading of dog training literature. But the experiences are always there, in the background, and they shape this text more than anything else.

Introduction: Canine-Human Intensifications, Periodizing Dog Training in the US Since the 1850s

> Today, through our ideologically loaded narratives of their lives, animals "hail" us to account for the regimes in which they and we must live. We "hail" them into our constructs of nature and culture, with major consequences of life and death, health and illness, longevity and extinction. We also live with each other in the flesh in ways not exhausted by our ideologies. Stories are much bigger than ideologies. In this is our hope.
>
> DONNA HARAWAY, *COMPANION SPECIES MANIFESTO*

•••

> [G]enealogy retrieves an indispensable restraint: it must record the singularity of events outside of any monotonous finality; it must seek them in the most unpromising places, in what we tend to feel is without history – in sentiments, love, conscience, instincts; it must be sensitive to their recurrence, not in order to trace the gradual curve of their evolution, but to isolate the different scenes where they engaged in different roles.
>
> MICHEL FOUCAULT, "NIETZSCHE, GENEALOGY, HISTORY"

∴

While this book provides a chronological look at the changes in the functions and methods of the training of companion dogs since the mid-nineteenth century onwards – on the basis of instructional dog training materials: training manuals, magazine articles and, as the twentieth century progresses, instructional videos and online materials – it aims to be much more than a chronology. It is meant as a genealogy – in the Foucauldian sense – of animal training; an analysis of the entanglement of North Americans' relationships with their pets in changing modalities of power; but also in debates about race, gender and species. The analysis fleshes out how training has been involved in the bigger

 | DOI:10.1163/9789004380295_002

cultural debates that have shaped and divided American society over a period of over 160 years: most obviously, the shifting role of companion animals in general, but also other – perhaps more surprising – issues; for example, concerning divergent attitudes to progress, scientific knowledge, folk beliefs, tensions between civilization and nature, reason and instinct, nature and nurture; changes in the gender coding of certain leisure activities and the methodological approaches within them. The genealogy of training can tell us more than how methods of making Rover sit have changed over time: it can tell us, if we probe deep enough, about shifts in the modes for the production of (human) subjects as well; it can tell us the story of consumerism; the ebbs and flows of the culture of sentiment in America and so much more. Dog training deserves to be examined through a comprehensive and complex approach because so far, its rendering has been very reductionist. If there exists a mainstream narrative of the development of dog training in North America – and some would claim there does not: the history of dog training is usually relegated to single chapters in books on dog behavior and articles in academic journals on applied animal behavior – it is a narrative of civilizational progress, according to which gentler techniques have developed as a result of our increased emotional sensitivity, fortunately – though somewhat coincidentally – backed up by contemporary scientific discoveries in animal cognition and behavior. It is a story of Enlightenment.

But this is not the only story. It is a dominant fiction, in the sense in which all histories are fictions because they rely on processes of selection and arrangement. Applying a genealogical method can help reveal how and why this particular story has been structured the way it has, while achieving the side effects of making it possible to both recover stories that have been deleted because they do not comply with the requirements of this narrative of progress (for example S.T. Hammond's methods, which just seem too progressive for the nineteenth century) or the dissenting counter-narratives (Vicki Hearne, Donald McCaig) that necessarily get lost in the epic narrative of the triumph of "positive training." Using the Foucauldian framework, which has been at times criticized as anthropocentric (Haraway 2008), but which has also been successfully extended onto the nonhuman world (e.g., C. Palmer 2001; Patton 2003; Pandian 2008; Shukin 2009; Nimmo 2010; Nealon 2016), makes it possible to detect patterns, while at the same time recovering the experiences of individual subjects living under specific regimes of power. It seems Foucault is beginning to be perceived as increasingly useful for animal studies, as evident in the recent proliferation of publications within animal studies that focus on the value of Foucault's method for the field, including the volume *Foucault and Animals* (Chrulew and Wadiwel 2016). In one of the first texts explicitly positing possible intersections between Foucault's work and animal studies, Clare Palmer asked: "Can Foucault's analysis of power be extended to include animals directly, that is, the ways in which humans and animals interact; their practices; their methods of attempting to control and discipline one another?" (C. Palmer 2001, 347). Many of the writers mentioned above answer with a cautious "Yes." However, while many of them casually mention training in relation to discipline, hardly anyone takes on training as a practice that is not stable but flexible and responsive to changes in modalities of power, but also as a practice that can be heterogenous when analyzed through a synchronic approach. And yet, in local bookstores Cesar Millan's books stand right next to those by Patricia McConnell, even though the two trainers represent incompatible discourses about what the canine-human relationship is and how it can be shaped. For me, the appeal of using Foucault is in his potential for conceptualizing both change over time and heterogeneity in a given time period.

However, I am more interested in the story of training than I am in Foucault: this is why at times this basically Foucauldian analysis gets messy. Firstly, I use Foucault much the way Foucault uses Nietzsche – whose *Genealogy of Morals* obviously serves as the inspiration for the title of this book and whose concept of genealogy inspired Foucault – that is, I deform him and make him "groan and protest" (Foucault 1980, 53). I do not see this book as contributing to Foucauldian scholarship; I see it as contributing to animal studies. There are times when rather than entering the debate on potential readings of a concept in Foucault's writings, I simply accept a Foucauldian scholar's reading because I find it useful for the argument I am making, which is an argument about dog training, not about Foucault. Examples include my use of the term "society of control," which I take not from Foucault himself but from Foucauldian scholars trying to conceptualize what happens after the waning of the disciplinary society; my hesitant acceptance of the distinction between repressive biopower and emancipatory biopolitics, which I take over from Michael Hardt's reading of Foucault (Hardt 2010), and which I apply in chapter 7 to theorize the most recent turn away from radical behaviorism in dog training. I am aware this distinction is by no means unequivocally accepted by Foucauldian scholars, but adopting it strengthens the argument and makes the shift more visible and easier to comprehend. On the other hand, Nikolas Rose's reading of the intersections of (human) eugenics and biopolitics (2009) is highly compelling and my reliance on Rose's reading of Foucauldian biopolitics in chapter 2 is reflected in my usage of the term "eugenic biopolitics" in a place where Hardt would more likely see biopower.

Secondly, I dive in to explore issues that are considered Foucault's blindspots: specifically, issues related to gender, class and race. Without these dimensions, the genealogical analysis would not be complete: the changes in the gender and racial coding of dog training are necessary for grasping the specificity of this activity in North America and its connection to major historical events in the US: chapter 1 suggests that the popular nineteenth-century practice of teaching pet dogs "tricks" that imitated human behaviors was enmeshed in anxieties related not only to species difference but also racial difference; species became a field for playing out racial tensions. Chapter 2 explores how the coding of the training of hunting dogs in the 1850s as an upper-class white male activity went hand in hand with restrictions on ownership of dogs by slaves in the US South and how the association of African Americans with certain types of dogs has served the purpose of stigmatizing both humans and dogs. Chapter 3 touches on a period of when dog training becomes a tool for the redefinition of masculinity; with the iconic scene of forcefully ensuring the canine's submission, of "dominating" the dog, serving as an allegory of man's conquest of nature, while chapters 4 and 5 show how the shift toward seeing training as carework tilted the gender balance among dog trainers. I spent my adolescence shuffling between the US and Europe. When I began to be involved in dog training, as I was growing up in Central Europe in the early 1990s, training (of dogs kept in the cities as pets – no less!) was firmly associated with working-class masculinity. In the same time period in North America, the field was already dominated by middle-class women, who were also, without exception, white (Włodarczyk 2016b). In fact, the market of companion dog training publications had been a vestige of whiteness in America (with an overrepresentation of German-American writers) until the phenomenal career of Cesar Millan, who – as chapter 6 argues – seems to have found a way of using his ethnicity to his benefit.

Adopting a genealogical framework makes it possible to challenge certain assumptions about the impulses for change within the field: particularly the story of gentling. The master narrative, at least among trainers themselves, is one of progressive (or sometimes abrupt, almost revolutionary) gentling of methods, as Donna Haraway somewhat jokingly put it, going from the dark ages of "leash jerks and ear pinches" to "the joys of rapidly delivering liver cookies" (Haraway 2003, 48). While the Foucauldian framework puts a twist on the notion of gentling – considering it not as an impulse of enlightened modernization but a turn to a more efficient modality of power – a close reading of the various textbooks makes it possible to additionally probe the narrative of gentling differently: it becomes clear that hardly any trainers ever advocated methods they considered cruel. It is popular understanding of what counts as gentle that has undergone redefinitions, and these redefinitions cannot be effectively described using an evolutionary scale. S.T Hammond's methods from the 1880s seem gentler than Cesar Millan's from the 2000s. Meanwhile, in the 1960s, William Koehler, a trainer infamous for the severity of the punishments he proposed, received an award from the American Humane Association, the same body which in 2006 issued a statement condemning the methods of "Dog Whisperer" Cesar Millan. Blanche Saunders, who taught the first obedience training class offered by the American Society for the Prevention of Cruelty to Animals, was once approached on the street by a passerby who threatened to report her to the A.S.P.C.A for "jerking the poor dog's neck" (Saunders 1974, 167). Philosopher Vicki Hearne was a huge advocate of trainers who are not so fondly remembered for the harshness of their methods, William and Richard Koehler, whom she read with very Nietzschean ideas about what constitutes moral behavior. Hearne argued that the Koehlers' harsh methods were infinitely kinder than the "humane and schizoid patter" of those who speak of kindness while effectively instigating the kind of permissiveness that swiftly leads a dog to being euthanized. Conversely, Ashley E. Pryor (2012) discusses kindness on the example of Cesar Millan, who has been accused of being potentially the most abusive dog trainer in recent years. These apparent paradoxes cannot be explained away with a story of gentling: a different approach is needed; one which recognizes "gentleness" as a slippery category. Foucault suggests in *Discipline and Punish* that the disciplinary system is necessarily based on the threat of penalty (Foucault 1995, 170–84). What really is a turning point in dog training is not the gradual "gentling" of methods but the rather sudden conceptual shift to the possibility of eliminating the notion of the penalty altogether. Whether this is feasible or not is absolutely secondary; what matters is that this idea – the idea that training does not have to resort to inflicting pain – emerges at a specific time, signifying the end of the disciplinary regime. This is what makes the behaviorist revolution in training so interesting (even though I read it as only a transition towards a more posthumanist approach): the foregoing of any kind of pain in training is its starting point. Not punishment – even though behaviorists might claim that it is punishment that their methods eliminate, their claim works only within a very artificial definition of what constitutes punishment – but their conscious desire to eliminate the choice between the painful and non-painful that in *The Birth of Biopolitics* Foucault, after David Hume, posits as an irreducible choice (Foucault 2008, 272).

Looking at methods within a time frame of over 150 years also helps to deconstruct the narrative promoted by opponents of training, that is the belief that any and all training is "antinatural domination made palatable by liver cookies" (Haraway 2008, 222). Such an approach, quite an arrogant one, overlooks the qualitative differences between the various methods and the changes in both the functions and the technologies of training over time. It is representative of the desire to dismiss the issue of training altogether, not of an intellectual curiosity to understand why and how certain cultural practices operate. The desire to dismiss the ethical conundrums generated by the issue of training was also discernable in the question posed by some colleagues who learned about this project: how do you separate training from socialization? In Kymlicka and Donaldson's *Zoopolis* (2011) the distinction between training and socialization signifies the border between morally acceptable and morally unacceptable interactions with companion animals. The honest answer is that I do not focus on defining this boundary. This book is not about providing answers for those who do not wish to become involved in this messy ethical field; it is an entryway into these complexities. While it is important to signal that increasingly more trainers, from all sides of the training spectrum, are becoming wary of the word training – and have taken to replacing it with substitutes such as "rehabilitation," "life skills" or "behavior modification" – it would not be fair to exclude them from a discussion of dog trainers simply on the basis of their desire to disengage from the tradition of institutionalized obedience training. The word training points to a rough field – more concrete than just human-canine interaction, though also always somewhat arbitrary – for studying the technologies of power exercised in human-canine relations, but as long as one is not a radical abolitionist who has never shared his house with a companion animal, it is hard to say one is immune from issues of training. However, there have existed humans who have chosen to become more engaged in training their canine companions, for reasons that have changed over the decades and that are explored in this book. Most generally, I read the training relationship as something of an intensification of the pet relationship: it is embedded in the same forces that shape the pet relationship, but it is more intense, more focused. While every canine guardian is – unavoidably – a trainer, those who explicitly choose to define their relationship with their non-human companions through the framework of training are also substantially different. Many dog guardians who currently, in the twenty-first century, embrace the label of trainer usually do so to emphasize their investment – of time, resources and emotions – in consciously shaping the human-canine relationship. They read their bonds with their dogs as more intense than those of "pet people." As performance artist and agility competitor Holly Hughes writes: "our dogs don't behave as pets; they refuse to stay on the periphery of our emotional life" (Hughes 2015, 212).

The genealogical method also puts an end to questions of the definition of dog training. When one reads the myriad definitions present in textbooks over a period of a century and a half, it becomes clear that the term training is a signifier that is attached to changing signifieds. There is a certain story that these signifieds tell, and it is this narrative that I want to investigate: from the discussion on training versus breaking that took place in the mid-nineteenth century and placed training (as opposed to breaking) within the framework of a civilizing endeavor and the Victorian ethics of kindness; through the highly technical behaviorist definitions that focus on changing behavior as a result of specific stimuli; via Vicki Hearne's work on the relational aspects of training; to contemporary definitions that move away from describing training as a unidirectional practice that changes the behavior of the dog to two-way practices that affect both the human and the animal who form a new entity as a result of the experience of training.

While the last definition does not come from a dog training manual but from Donna Haraway's *When Species Meet* (2008), it is an empirical fact that the majority of rulebooks for contemporary human-canine activities prioritize the relationship-building aspect of the practice over the usefulness of the activity for the human partner. In fact, the affective value of the companion animal is increasingly officially recognized as the new use value, as proved by the recent emergence of Emotional Support Animals: animals that are seen as playing a therapeutic role not because of any useful service rendered to their owners but simply because of their presence; or their affective labor, depending on one's perspective. Training had long been tied to increasing an animal's use value: a fact recognized within the growing field of animal-assisted therapy. Therapy dogs are required to be trained (and certified) in order to be considered of use. The recent debate about Emotional Support Animals who – unlike service dogs – do not have to be trained, brings to the surface tensions related to the definition of use value as it relates to companion dogs. Finally, I argue in the last chapter that some contemporary training practices can be conceptualized through the human's desire to become more like the dog, in an absolute reversal of the nineteenth-century definition of training as a procedure that "humanizes" the canine. To conceptualize this shift, I turn to Sergei Prozorov's recent work on the potential for finding an "affirmative biopolitics" in the late writings of Foucault (Prozorov 2015).

What echoes in the various historical definitions of training is the "nature vs. nurture" debate as it pertains to canines. Training is often viewed as the "culture" to the dog's "nature" and thus, depending on one's view of culture, it can either be seen as a stifling practice of cultural domination or as a guiding principle for peaceful interspecies coexistence. However, especially in the case of companion animals, this is a simplified understanding. Ethologists Raymond and Lorna Coppinger speak of trainability as the distinguishing feature of dogs: that which makes them different from wolves (Coppinger and Coppinger 2001, 47–49); an observation reiterated from a different disciplinary perspective by Serpell and Hsu (2005). The Coppingers' observation suggests that trainability is part of the legacy of human-canine long-term coexistence: dogs would not have developed into dogs if humans did not attempt to train them and did not use trainability as a criterion of breeding selection. Consequently, training has changed the behavioral conformation of dogs in a way that is passed on to subsequent generations. This understanding of training is by no means novel. Darwin's disciple, George Romanes, wrote of dogs' "natural sagacity" as augmented and developed by "persistent contact with man, coupled with breeding and training" (Romanes 1882, 437). Offspring of outstanding sheepdogs have a greater likelihood of becoming successfully trained as sheepdogs themselves. At the same time, training is needed in order for the genetic potential to be placed under stimulus control. As such, training is – or at least has been – enmeshed in the framework of eugenic biopolitics. At the same time, the behaviors that are viewed as "natural": the eyeing of stock by a sheepdog or the pointing behavior of a gundog have also been shaped by certain purely cultural preferences. Of course, there is a huge difference between balancing a dog biscuit on the nose and bringing in a flock of sheep, and while this book is concerned primarily with the training of pet dogs (biscuit on the nose), the relationship of trainability and heritability is one of the tensions that is taken up, especially in the discussion of nineteenth-century literature. Tying Foucauldian discipline into the discussion, Clare Palmer notices that: "[i]n this context, animal discipline and nonresistance is achieved by inscribing on the very genetic make-up of the animal body preferred physical and behavioral characteristics" (C. Palmer 2001, 355). Yet what happens if we redefine training as possible outside of the disciplinary regime? What if the goal of training is no longer conceptualized as discipline and nonresistance: the production of a docile body? How does that change the relationship of training to "eugenic biopolitics"?

One could argue that dog training is a field of inquiry like many others and that similar observations about the intricate and multi-dimensional workings of power could be made using different examples; the education and upbringing of children being an obvious, though most certainly not the only possible, area of study. Foucault wrote about the emergence of specific institutions: the prison, the school, the hospital, the mental asylum, as turning points in the development of disciplinary power. From this perspective, the dog training school is yet another such institution. While this perspective is not false, it is not exhaustive. Dog training, in addition to being a field that I have been intimately familiar with, is not just a microcosm of the powers present in society in general. Dog training is not just representative. It is also unique because it is a field of human-nonhuman interactions. Dogs are not going to tell us their stories, at least not in the most literal sense of the word: their stories have to be recovered differently from the stories of humans who were able to leave behind written (or otherwise linguistic) accounts of their lives. While this book is not an attempt to write from the animal's point of view, to allude to the title of Eric Baratay's work *Le point de vue animal* (2012), it is an attempt at what Erica Fudge calls a holistic history (Fudge 2002): it shows how the stories of humans and non-humans are entangled and shaped by the same broader forces.

The human-canine bond is not just embedded in modalities of power: it is also a performance of the key concepts structuring these modalities. By watching dog-human interactions one can say a lot about the prevailing social values of the time and, by extension, looking at how these structured interactions change over time is one way of evaluating social change. The performances cannot be evaluated outside of the material context of their production. It would be a gross oversimplification to look just at the finished product: the trained dog. It does matter how the dog's performance was elicited, reinforced, what tools were used, what behaviors the performance consists of, where it takes place, who watches the act and why, etc. This is why this project is not based just on an analysis of the performances themselves but of the entire apparatus that went into their making, as presented in the instructional materials that serve as primary sources. The behaviors taught to dogs are telling and so are the fashions and fads for teaching certain behaviors and not others. However, a trick in which a dog walks on his hind legs was an element of a very different apparatus in the mid-nineteenth century than it is in the early twenty-first. Similarly, the image of an emaciated dog pulling a milk cart on the streets of early twentieth-century Montreal is part of a different apparatus than a contemporary image taken in the suburbs of Chicago and featuring a purebred Great Pyrenees hooked up to a fancy pulka cart imported from Norway.

1 Periodizing Dog Training with Foucault

The very concept of training, understood as the purposeful modification of an animal's behavior, could not be conceptualized within the regime of sovereign power, which relied on spectacle as its primary practice. The purpose of the spectacle – the example analyzed by Foucault in the opening of *Discipline and Punish* being Damiens' execution – was to deter potential criminals from committing a crime by making them fear the fate of the unfortunate individual whose body was directly marked and mutilated. Obviously, the effects of such a spectacle on a non-human audience would be limited, to say the least.[1] Sovereign power assumes an understanding that extends beyond the spectacle itself: a connection needs to be made by the spectator between the past transgression, the resulting – though not necessarily immediate – punishment and the possible future fate of the spectator. This is an understanding that simply would not arise in the case of dogs, even if they are – as recent studies prove – capable of learning through imitation and social interaction. This explains why animal training can only be conceived of as beginning with the end of sovereign power: in a framework that assumes the possibility of change, improvement of behavior and one which does not work through spectacle. It is not coincidental that as the training of pet dogs truly gains momentum in the 1850s, displays of cruelty to animals are already viewed as brutal and uncivilized (Mackintosh 2016). It is not surprising that training is actually the word of choice for Foucault when discussing the production of human docile bodies in the disciplinary regime. In fact, the French original uses the word *dresser*, most often applied in the context of animal training and recognizable in the English-speaking world through its association with equine dressage. In other words, training is born out of improvements in the efficiency of the functioning of power, for let us not forget that Foucault's account of the different modalities of power is hardly a story of progress or Enlightenment, but a story of what Jeffrey Nealon calls "the increasing 'intensity' (*intensité*) of power: which is to say its increasing 'lightness' and concomitant 'economic' viability" (Nealon 2008, 32). Modernization is thus not about greater gentleness; it is about improved efficiency. Gentleness is something of a byproduct. The subject whose behavior is to be changed, cannot be killed in the process; this would be pointless from multiple perspectives, though most significantly from the economic one.

2 1850–1910: Shaping the Dog's Soul

Foucault's disciplinary power – often casually mentioned in the context of animal training – does not suffice to explain the undertaking of dog training in the period that this book opens with, that is in the second half of the nineteenth century. When we conceptualize humans' relations with non-human animals, periodization – as applied to the humans' relations with other humans – tends to break down. For example, to this day eugenics remains the predominant paradigm for canine reproduction even though it has been

1 This is, of course, not to deny the by now well-researched historical practice of public executions of animals.

viewed as unredeemably corrupt in the human world for over seventy years. The nineteenth century is something of a transitional period in animal training, one that could be rendered using the title of a rarely analyzed chapter of *Discipline and Punish*: "The Gentle Way in Punishment," in which Foucault discusses a transitional modality between sovereign power and disciplinary power: social power. In this chapter, Foucault introduces his guiding thesis for the entire book: the claim that the turn towards gentler methods of punishment, from public execution to prison sentences, is not about modern society becoming more civilized (despite this being the common-sense understanding) but about a different logic of social organization; with modern society being controlled by the middle class with different social agendas than the nobility. This period of "social power" is transitional in that it sees the need to move away from the old style of punishment and yet it has not worked out the efficient methods that will provide the nuts and bolts of the disciplinary mode. As such, it is interesting because it is still full of inconsistencies; tears that reveal the confusion of the transitional moment. Dog training here becomes something of a microcosm of the entire power apparatus; it epitomizes, on a smaller scale, the changing modalities of punishment in a modern(izing) society.

Training is never about purely eliciting behaviors, but this is never more evident than in the second half of the nineteenth century; the period when training textbooks emerge. The books from this period are not so much concerned with methods as they are with firmly embedding the practice of training in respectable, bourgeois pursuits, and the very emergence of a market for such books in the United States is also tied to the changing functions of pet-keeping in an increasingly urban society. In a period of increased industrialization and urbanization, interactions with nature were perceived as having a beneficial effect on humans, particularly on the shaping of young people's character. This tenet of Victorian[2] society, including the perceived link between cruelty to animals and corruption of one's character, has been researched extensively by animal studies scholars (Ritvo 1987; Kean 1998; Grier 2006). At the time, the discourse of dog training attempted to become imbued with the rhetoric of a civilizing endeavor, akin to "the white man's burden" of "civilizing savages," while remaining firmly embedded in the concept of human stewardship of nature and partaking in the Victorian "ethics of kindness" to all creation. The place of training within other respectable activities was not a given; it had to

2 The word Victorian is used very broadly here: also to refer to American middle-class, bourgeois culture of the nineteenth century. I am, of course, aware of the specificity of the American context, yet whenever I used the term Victorian to refer to the US I am alluding to the similarities in the ideological projects of the English-speaking world in the nineteenth century.

be secured through skillful rhetorical maneuvers and continued to be under threat in light of the association of trained dogs with low-brow entertainment, such as the dog and pony show and circus acts (Davis 2002; Young 2015). The public had to be convinced that the beneficial character of the process of training for a pet dog was to be located in the fact that training elevated the animal subject by making him endowed with certain almost human-like skills and brought to the surface the sometimes deeply buried positive qualities of a dog's character. Thus, instead of detailed instructions, the textbooks are often full of stories of loyal and "sagacious" (trained) canines, who help their human masters in times of need, whether by going grocery shopping or by sacrificing their own lives for those of their masters. It is the training that the dogs have received that supposedly helps them transcend their "brutish" behavior. At the same time, as chapter 1 discusses, a close analysis of the entire training apparatus reveals how it re-established the human-animal boundary, which the rhetoric of training in the Victorian age seemed to challenge, constantly reminding the readers that dogs can never be as moral and intelligent as humans.

In Foucault's reading, social power was supposed to operate on the soul of the individual: opening up the possibility of reform and imbuing punishment with a deeper life-transforming dimension. The training manuals of the period certainly present dog training as engaged in this bigger project of both social and individual reform: while the issue of animals literally having souls is not discussed overtly in the textbooks, training is shown as benefiting the dog's mind and improving his status in the hierarchy of creation. As Francis Butler, the author of *Breeding, Training, Management and Diseases of Dogs*, wrote in 1857: "My object is to prove that it requires but a small amount either of time or labor, to rescue even the meanest mongrel from a life of degradation and misery and place him in a position, where he may be respected for his virtues and appreciated for his worth" (F. Butler 1877, 88). Within such a modality of power, punishment takes on a range of different, often almost contradictory meanings. On the one hand, it clearly has transformative potential; it is only through punishment that this supposedly higher state of mind can be achieved. On the other hand, the punishment cannot be of a too drastic nature. Brutal physical punishment was deemed uncivilized and having a corrupting effect both on the trainer and on the animal, not to mention that it plainly contradicted the Victorian ethics of kindness. To resolve this contradiction – the belief that punishment is necessary because of its transformative power with the belief that excessive cruelty has no place in a civilized society – nineteenth-century trainers created an intricate punishment apparatus, which approved of only certain tools and certain ways of administering corrective measures. Punishing a dog in a fit of rage, an outburst of emotion, was not acceptable. Of the tools for administering punishment, kicking was deemed the most offensive, while the whip was seen as the tool of choice for effective punishment delivered in a calm manner. Frank Forester wrote in 1856 that to kick a dog is "an act of utter and unpardonable brutality" (160) while "a slight flogging, with a great deal of talk, remonstrance and scolding, has much more effect than a savage, passionate beating" (157–58). Such advice generally corresponds with the Victorian belief that savagery is associated with instinctiveness. It is civilized to temper one's instincts, whether one is human or canine.

At the same time debates about what instruments are proper for administering punishment – the whip was the tool of choice for almost all trainers, but heated debates arose regarding, e.g., the merits of spiked (prong) collars versus choke collars – reveal the absolute arbitrariness of the notion of "the gentle way in punishment," at least as regards dog training in this period. What the preoccupation with gentleness implies is that it was desirable to be considered gentle, as this was an indicator of the trainer's civilizational progress and social status. In other words, nineteenth-century trainers were primarily invested in proving that their training practices were a sign of their own benevolence and progressiveness but were cognitively unable to conceptualize the process of training as not based on punishment. Susan Nance makes a similar argument as regards the training of circus elephants in the period under discussion, positing that claims of gentleness were more of a public relations strategy than an actual change of methods (Nance 2013, 72).

Chapter 2 looks at the different functions that training served in the world of working dogs in the second half of the nineteenth century, using the example of hunting dog manuals. These prove that training was a tool of eugenic biopolitical selection: trainability was a significant criterion in the decision-making process of the breeder of working dogs. Yet, it was the Victorian culture of the period that defined the notion of trainability. In the circle of upper-class sportsmen, trainability referred to the dog's ability to control his own canine urges, his impulsiveness. Dogs who were just as effective as hunters but hunted in a style that did not emphasize self-control not only had less prestige but became associated, particularly in the antebellum South, with African Americans. Claire Jean Kim, in *Dangerous Crossings* (2015), notes that "[a]nimalization has been central not incidental to the project of racialization" (18), and the story of slaves' hunting dogs in the antebellum South reveals the mechanisms of this process. As racial tensions mounted, legislation prohibiting the ownership of dogs by slaves was implemented: using the dogs' supposed lack of trainability (and the danger they posed for the landowners' sheep because of it) as justification of their killing. The fate of dogs owned by Virginia slaves in the 1850s can serve as a departure point for reflection on how trainability can be implicated not only in "making live and letting die" but also in "making die." In a legacy that haunts the dog world to this day – particularly in contemporary stories associated with so-called pit bulls and other breed-specific legislation (BSL) – dogs deemed untrainable have been, and sometimes still are, killed. Yet the definition of "trainability" is entangled in a framework that is racist and classist at its core.

3 1910–1970s: The Emergence and Strengthening of the Disciplinary Regime

The early twentieth century is the time of the emergence of disciplinary institutions connected with the training of working dogs; most significantly military dogs, police dogs and guide dogs. Europe led the way in establishing such programs and codifying methods, which were then adopted also in the USA. The first American school training guide dogs for the blind, The Seeing Eye, was established in 1929 by Dorothy Harrison Eustis, who had previously lived in Switzerland and bred German shepherds there. The US was a latecomer in the training of military dogs; entering the field through a rather odd program that linked the world of pet dogs with the world of working dogs: Dogs for Defense. The campaign by this name, instigated immediately after the Pearl Harbor bombings, promoted citizens' vicarious participation in the war effort through donations of family pets to the military (Downey 1955). Forty thousand pet dogs were donated and 8,000 of those successfully completed training and were deployed (Alger and Alger 2014, 84). After the war, the dogs were "retrained" for civilian life and a number of those who were able to complete the retraining process were returned to their families of origin.

The Dogs for Defense campaign is representative of the disciplinary management of canine bodies in training practices in the first half of the twentieth century. Firstly, training becomes the province of institutions organized specifically for the purpose: the transfer of a canine's body from one space to another marks visible stages in the training process. Protocols begin to be established in order to evaluate the success of a training program: standardized examinations become the primary form of evaluation. From the ethical perspective, the sacrifice of canine bodies in lieu of human bodies in war is not seen as a questionable endeavor, as evident in the public's enthusiastic response to the call for canine soldiers. In fact, and this could be termed the trickle-down effect of the military mindset onto the pet-owning population, a dog's usefulness to the larger human community – understood, to a certain extent, as the willing sacrifice of a dog's canine happiness – becomes an obvious mark of distinction. The discourse of training is no longer about providing wholesome entertainment for humans while developing a dog's intellectual prowess; it is about proving the dog's usefulness to humankind. And it is serious business. There is no expectation of canine joy: training is the meticulous and detailed inculcation of the canine into his social role as the faithful servant of man. The great irony of this enterprise is that it is undertaken at a time when increasingly fewer dogs are actually used as helpers in activities providing basic sustenance for humans: hunting has become largely a sport, herding is on the decline. It is the uniformed professions, particularly the army and the police, which are creating new opportunities for, so to speak, canine employment and which also provide a model for organizing the training of pet dogs.

Even though Konrad Most's *Dog Training: A Manual*, a book widely dubbed the first modern book on dog training was only published in the US in 1955, Most's disciples popularized his methods and exercises among American canine enthusiasts. Most's approach symbolizes the birth of obedience, not as a competitive event (two women, Helen Whitehouse Walker and Blanche Saunders, were instrumental in establishing obedience trials in the US) but as a conceptual approach to training: one which emphasizes the dog's submission to the human. In the nineteenth century, the trained dog was one whose faculties of reason were seen as greatly improved as a result of training: the sagacious dog. In the early twentieth century, the trained dog is unconditionally obedient and not necessarily smart: Most is credited with emphasizing the distinction between intelligence and trainability. The enthusiastic welcome given to Most, his methods and his exercises was not so much an issue of their practical value as of their helpfulness in securing social prestige for the activity of training and of their timeliness, specifically their compatibility with changing views of nature and masculinity. As previously mentioned, nineteenth-century trainers' attempts to popularize training as a respectable middle-class pursuit were hindered by the association of trained dogs with circuses. The training books of the Victorian period routinely discussed the teaching of tricks. Yet, tricks connoted leisure, uselessness, even a certain decadence and effeminacy. In the age of "Teddy Bear Patriarchy," to use the title of Donna Haraway's article on the intersections of gender and nature in the early twentieth century (1984), the military drill of Most countered the effeminate decadence of the circus and the association of obedience training with the training of army dogs secured the much-needed moral and political dimension of the project. After all, dogs were being prepared to serve their master, and – by extension – their nation, at least potentially. As Jennifer Mason argues in *Civilized Creatures*, the perception of domestic animals began to change in the first decades of the twentieth century, and it is their latent wildness rather than their ability to lead a civilized existence that came to be valued (Mason 2005, 159). It is not coincidental that a new model of masculinity gains significance towards the end of the nineteenth century. In a recent analysis of the role of dogs in sentimental literature, Małgorzata Rutkowska adds that "shifts in the cultural perception of masculinity influenced the way middle-class men wished to assert their identity. An ideal Victorian gentleman was cultivated and restrained, yet this self-control now came to be seen as repression of natural masculinity manifested in powerful, primitive instincts" (Rutkowska 2016, 46).

Concurrently, the performance of the human-canine bond becomes a re-staging of the primary scene of man's conquest of nature. The dog needs to be physically subjugated in a scene that speaks of the trainer's courage and bravery in the dangerous undertaking of dominating and taming the wild beast. The dog is both an object to be controlled and a source of vitalistic, animal energy that can be appropriated by the trainer in the process of training. The military drill of Konrad Most and his North American followers is usually looked upon with disdain by contemporary trainers for whom those are "the dark ages" of dog training: the times of choke chains and prong collars. Yet, as chapter 3 proves, when the methods themselves are analyzed in detail, they are not more physically intrusive than those from the books from the preceding period (1850–1910). What changes is the discourse: rather than speaking of gentleness, benevolence and patience, the books list firmness, strength and courage as the required personality traits of the trainer. The process of training the dog, for the first time, becomes a physical struggle, much like the taming of a wild animal. The confrontation of wills is seen as an unavoidable element, and it is the breaking of the dog's desire for freedom that proves the success of training, though constant vigilance is required in case the dog decides to revert to the "state of nature." Yet, the confrontational methods that emerge in the twentieth century, and especially the infamous "theory of dominance" – that is the mistaken belief that the human-canine relationship needs to mirror the hierarchy in a wolf pack and that the dog needs to be physically subdued to prevent him from attempting to take over the alpha position – were actually part of the attempt to make the training relationship more "natural": with teaching techniques mirroring those found "among animals themselves." Anthropomorphism, so typical of Victorian culture, was recognized as fallacious. The Victorian contempt for animality – understood as the inability to control one's urges, as instinctive behavior – becomes replaced with an almost Nietzschean fascination with animality. For Nietzsche, the animal is a figure toward which the human should strive precisely because animals do not repress their natural urges and, as Claire Rasmussen writes, they "lack the kind of self-limiting tendencies in humans that Nietzsche sees as causing weakness" (Rasmussen 2011,

107). While the dog is not Nietzsche's example of animal "will to power," the process of subjugating the dog and enforcing his obedience most certainly is.

Interestingly, the emergence of institutions for the training of working dogs in the United States coincides in time with the codification of an organized past-time modeled after the training exercises invented for police dogs and hunting dogs: dog obedience. Again, somewhat ironically, the persons to whom credit is due for promoting this activity in the United States were neither police nor hunters. Helen Whitehouse Walker and Blanche Saunders, whose colorful adventures featuring poodles and one of America's first recreational vehicles are described in chapter 4, pressured the American Kennel Club to add obedience trials to the list of sanctioned events for purebred dogs. Their own rather inconspicuously hidden agenda focused on the rehabilitation of the poodle as a working breed. By the early twentieth century, canine biopolitics had become refocused: dog shows gained prominence as the primary tool of breeding selection. Poodles quickly became one of the most commonly exhibited breeds and came to be known for their looks rather than intelligence. Walker initially saw obedience training and obedience trials as a possible tool for breeders who wanted to "preserve the intelligence of their breeding stock." Obedience trials were designed as an attempt to return to the old biopolitical frame of selective mating of dogs whose field performance was outstanding. Training classes, not to mention the trials themselves, were restricted to purebred animals (AKC trials opened to mixed breed dogs only in the twenty-first century!). Yet, the attempt failed, as it became clear that the capability of learning obedience exercises was not – at least not clearly and unequivocally – heritable.

The American dog writers of that period, some of whom were former students of Most, argue that what American pet owners need is to teach their dogs how to be companions, not guardians. It is with the emergence of these ideas that dog training begins to be involved in the project of governmentality: training becomes a tool for managing the behavior of the urban canine population on a mass scale, in a way that renders dogs docile in a very literal way: they will not bite or attack humans, nor interfere in the daily lives of the non-dog-owning population. While the function of training is increasingly related to governmentality, the methods are typically disciplinary: efficiency is key; the methods for teaching the equally uniform exercises are homogenous and can be applied to any dog; training takes place in a group context and its success is evaluated through a formal examination. Because training is meant to address the behavior of every dog, its effectiveness ceases to serve as a criterion for breeding selection. Chapter 4 also analyzes the changes in gender coding that the activity of companion dog training underwent in this period, viewing them not just as an individual achievement of the women involved in the training field but as the beginning of a bigger paradigm shift.

The uniqueness of the dog training school as a disciplinary institution consisted of the fact that the subject of discipline was not just the dog but the human-canine dyad. As echoed in the title of Blanche Saunders's most famous book, *Training You to Train Your Dog* (1946), the instructor taught the human owner how to train the dog. Resultingly, though as something of an unintentional side effect, the bond between the dog and the trainer strengthened to the extent that some trainers began feeling uncomfortable using the training methods advocated by the instructors. The rift between the goals of governmentality – achieving peaceful human-animal coexistence – and the technologies of discipline used to achieve that goal led to increasing criticism of the disciplinary model.

4 1980s–2000s: From Governmentality to Self-Governmentality: Biopower, Behaviorism and Care of Self

Yet, it is only with the rise of the prominence of behaviorism in training methods in the last decades of the twentieth century that the link between breeding and training of companion dogs becomes completely severed, and training begins to be defined as the process of human-canine relationship-building. While this may seem surprising, bearing in mind Skinner's bad press among contemporary psychologists, behaviorism finally provided a technology that made it possible to conceptualize training without pain and punishment (at least if one defined punishment using a Skinnerian vocabulary); a necessary step in the move away from discipline. B. F. Skinner's insistence that any organism can be taught any behavior that it is physically capable of performing (Skinner 1938), and the detailed protocols for eliciting and shaping behaviors, redefined training from a somewhat vague and mysterious process of capitalizing on the animal's instincts and inborn talents to a scientifically-governed procedure with easily repeatable effects. Most significantly, it is only with the behaviorist understanding that training the household pet begins to make sense for the average pet owner. If proper behavior is not inherited (as Skinner insisted), it needs to be taught. Skinner's work had a double impact on dog training: not only did it prove significant for popularizing the notion that every dog needs training but, in yet another ironic twist of fate, B. F. Skinner, who during World War II led a project in which pigeons were trained as suicide bombers, became a key figure for the most significant revolution in dog training and one motivated primarily by ethical concerns: the turn away from reliance on aversion and toward positive reinforcement. I see this turn as motivated by what Nicole Shukin calls "feeling power" (Shukin 2013), a fantasy of mutual human-canine interspecies love and devotion. Yet the application of Skinner's behaviorism deconstructs the fantasy of feeling power, creating a technology of love: a set of techniques that makes is possible to manipulate the dog's affects.

It is also in the second half of the twentieth century that training ceases to be an element of eugenic biopolitics in the most apparent sense of the term: at least in the case of companion dogs, it is no longer a tool for managing the reproduction of canine populations. However, this does not mean that dog training does not have a place in the biopolitical order of the late twentieth and twenty-first centuries. Training increasingly becomes a strategy for managing the behavior of the canine population as a whole, of ensuring that it is not a threat to the human population; training becomes a tool of governmentality. In this sense, it is used as an instrument benefiting the human population. But something else begins to happen toward the end of the twentieth century: training also becomes a part of the set of practices aimed at shaping canine life, especially its quality and, by extension, length. Training begins to visibly benefit the quality of life of the canine population. There exists extensive contemporary research proving that the more training dogs receive, the less likely they are to be abandoned (Salman et al. 1998). Conversely, training programs for dogs in shelters increase adoption rates and the rates of retention after adoption (Luescher and Medlock 2009, Siettou et al. 2014).[3] Frederic Gros, a Foucauldian scholar who reads biopolitics as intimately entangled with

3 The results of the 1998 study were striking: 96% of dogs relinquished at animal shelters had never attended obedience training classes, 98.8% had never been in the care of a professional trainer (classes, private lessons, behavioral consultations). These results led to the establishment of dog training programs in shelters themselves. While research on the effectiveness of these programs is relatively optimistic, that is it proves that such programs do help increase adoption rates, their success (usually evaluated by the retention rate) is not as staggering as had been assumed. This suggests that the relationship building aspect of training is of key importance for the retention of pet dogs by adopters. This view is actually compatible with my reading of contemporary training as moving outside of the realm of pure discipline. In the disciplinary model, the trained dog is a product of an institution. The dog's relationship with the human being is secondary to the process of training: the effects of training are the dog's ability to respond to obedience commands. Training has now become a strategy for creating a strong affective relationship with the dog and such a relationship must be built by the new guardian him- or herself. The fact that the newly adopted dog knows how to sit is not a predictor of the strength of the potential relationship. The fact that the dog engages the human in play is such a predictor and recent studies show that the dog's openness to interactions with humans boosts adoption rates more than the dog's knowledge of obedience commands (cf. Protopopova and Wynne 2014).

neoliberalism, argues that the biopolitical regime is governed by notions of profitability. Education, as Gros discusses, had long been about "learning civic values" yet in the neoliberal/biocapitalist regime it is "about increasing the value of capital" (Gros 2016, 270). If, after Donna Haraway (2008) and Nicole Shukin (2009), we think of animals as biocapital, then training in the disciplinary regime turned canines into valuable members of society, at the same time making canine bodies disposable in the name of the greater good of the nation. Training in the biopolitical regime is akin to the human making an affective investment in the dog, thus increasing the affective value of said biocapital and strengthening the human-canine relationship. Roughly stated, the dog's situation from the point of view of his welfare is better under contemporary biopolitics than under discipline not only because he is less likely to be physically mistreated, but also because he is less likely to be disposed of or sacrificed.

Foucauldian biopolitics, albeit a concept that was never fully developed during Foucault's lifetime and one whose coherence remains disputed, is generally understood to consist of two poles. One is focused on the individual body, on maximizing its forces and integrating it into efficient systems and the other one on regulatory control on the level of the population, the species body, birth, morbidity and longevity (Rabinow and Rose 2006, 196; Cisney and Morar 2016, 4). Within animal studies the second pole has dominated academic discussion and has been useful for the analysis of the exploitation of animal bodies within the industrial capitalist system (including factory farming, the breeding of non-companion and companion animals, etc.) and for understanding species politics (primarily extinction, preservation, conservation, the notion of invasive species, etc.). In fact, the second pole is also dominant in the discussion of biopolitics as related to humans, although there are scholars interested in how the individual body becomes integrated into efficient systems of production through postdisciplinary techniques. Those who do focus on the first pole often explore the relations between the broader concept of biopolitics and the technologies of subjectivation within this regime, pointing to the renewed importance of Foucault's notions of technologies of the self and of pastoral power (Rose 2009).

In the case of dog training, analyzing how the technologies of subjectivation function in the biopolitical regime proves highly fruitful because the shift between discipline and biopower is potentially even more dramatic than in the case of human subjects: extreme physicality is involved in the disciplinary process of dog training, while the postdisciplinary techniques turn away from the body. The disciplinary regime addresses the body directly through training techniques aimed at the body: the molding of the canine body through human touch (pushing down on the dog's back to make the dog sit, pulling down the front legs or putting pressure on the withers to make the dog go down) or multiple tools: collars, leashes, training lines. The disciplinary premise is that if the body complies, the mind will follow. The dog will learn to associate the positions of his body with visual and sound stimuli provided by the trainer. The dog training school, as a place where the protocols of disciplinary training are implemented, is the pinnacle of the disciplinary regime. The type of positive training that emerges towards the end of the twentieth century, most strongly represented in clicker training and based on Skinner's operant conditioning, is a completely hands-off type of training. There is no touching of the dog, no molding of the canine body. What is targeted is the dog's mind through the process of developing the dog's motivation under the premise that the dog's body will follow the mind. Through meticulously prepared training protocols, the dog learns to want to be with the human and to willingly participate in the proposed activities.

Deleuze, in his take on Foucauldian biopower, lists the factory as the iconic space of discipline and the corporation as the best metaphor for this new type of arrangement, which he calls "the society of control" (Deleuze 1992), a term that is later taken up by Nikolas Rose in *Powers of Freedom* (1999a) and Hardt and Negri in *Empire* (2001). A work ethics of motivation permeates the corporation, but control does not disappear. On the contrary, it becomes so pervasive that it crosses the physical boundaries of the workplace which cease to matter (work from home, constant availability via cell phone, etc.). Similarly, in dog training, the emphasis on motivation does not preclude human control of the process of training. If anything, the control is even greater and extends to the minutest spheres of a dog's life. Training is no longer something that takes place at dog school: it becomes homework, and it never ends.[4] An interesting, so to speak, by-product of this shift is the active role the dog begins to play in the process, despite the human's intense supervision. It is the dog who chooses to engage (no leash) the human in joint activities. The key exercise, as analyzed in detail in chapter 5, is so-called "self-control": a set-up in which the dog must learn to control his own impulses first, in order to satisfy them

4 One of the primary complaints expressed by trainers with whom I interacted with over a period of many years is that people who come to training classes do not understand that in order to achieve results, they must train at home. This is not a complaint about the laziness of the pet-owning population but an expression of a fundamental clash of power regimes. The expectations of those who come to classes are a carryover from a regime where discipline was produced through the work of institutions under the guidance of experts. The trainers, often more progressive than their clients, define training through its relationship-building aspects.

later. The dog begins to be seen as capable of self-governmentality. Paraphrasing trainer Susan Garrett, the dog learns that to get the cookie, he must first give up the cookie. Much training becomes about generalizing the skill of self-control onto various spheres of the dog's life. When the description of such training techniques is read alongside the definition of the technologies of the self, a term Foucault introduces in Volume II of the *History of Sexuality: Use of Pleasure*, the similarity between them cannot be overlooked. Even though he is clearly a subject of power, the dog is invited to work on his own self, an invitation he may – at least theoretically – take up or reject. A somewhat similar theoretical argument is made by Anand Pandian, in an article about pastoral care of oxen in India: the oxen are expected to transcend their animality by not acting in ways dictated by instinct but by following a certain human moral code. Those who are the most willing to comply with the human moral code are rewarded with the greatest esteem and respect of their caretakers. Those who do not willingly comply are sent to slaughter (Pandian 2008). In the case of dog training, the consequences are usually not as dire; dogs are hardly ever abandoned by committed trainers because of their "animal nature." However, it is those dogs who exhibit the most motivation for playing the game of training that are valued the most highly by trainers.

Even though in the disciplinary regime power was represented in such figures of authority as the teacher or the physician, the goal was such docility that functioned even in the absence of individualized representations of power, as exemplified in the Panopticon. In dog training, the emphasis on the trainer-dog relationship was only cursory because training equaled a trained canine body, which performed regardless of the handler. The types of techniques that emerge after disciplinarity stand in stark contrast and shift emphasis onto the dog-human bond, which becomes a relationship of bilateral affection. The affection for the human is created in the dog, at least to some extent, through the techniques described above. The human's affection is often a starting point, but it is also strengthened via training. Such power that creates a dynamic and strong bond between the one who acts (in good faith) and the one who is acted upon is described by Foucault as pastoral power. As Pandian summarizes Foucault, pastoral power was "a form of power that was both individualizing and totalizing, carefully ensuring the distinct and particular welfare of each one within a population of subjects, while at the same time submitting all of them to greater degrees of discipline and control" (Pandian 2008, 89). While pastoral power is not ethically unproblematic, particularly as applied across the species line, its appearance also marks an undeniable turning point in the annals of the human-canine relationship. Not only does training begin to be pleasurable for the dog, even if the dog's desires are shaped through skillfully applied training techniques, it also begins to address the dog as an ethical subject with certain (albeit limited) decision making capabilities. Finally, true pastoral power brings with it notions of responsibility: the shepherd never abandons his flock and must be ready to sacrifice everything for the good of even the individual sheep.

5 2000–2015: Beyond Behaviorism: Affirmative Biopolitics

The last chapter of the book, chapter 7, explores the recent disenchantment with radical behaviorism in the dog training world and ties it to changes in contemporary attitudes to anthropocentrism: the belief in the centrality of the human in theoretical accounts of the world. In fact, dog training also begins to experience what has become known in the academy as "the animal turn": an ethically-motivated preoccupation with animal alterity. While this is not uniformly true of all dog trainers, a certain trend begins to emerge. Trainers cease to view their dogs as either bodies that must be disciplined or subjects whose lives need to be closely monitored and in whom a certain human moral code needs to be inculcated. Most importantly, the dog's "animality," rather than being something than needs to be repressed in the process of training, becomes a desirable quality. This redefinition of training is succinctly summed up in the title of Michał Pręgowski's article that discusses the shift away from radical behaviorism in contemporary dog training: "Your Dog is Your Teacher" (Pręgowski 2015). Some theorists have recently begun to re-read Foucault's work on biopolitics using Michael Hardt's distinction between repressive biopower and emancipatory biopolitics (Hardt 2010). They argue that an affirmative biopolitics can be located in Foucault's later work and that what accounts for its emancipatory potential is specifically its affirmation of animality (Prozorov 2015).

The previously underanalyzed significance of animality in Foucauldian biopolitics is also crucial for Jeffrey Nealon's argument about the emergence of animal studies (2016). In response to multiple accusations from animal studies scholars about Foucault's omission of animals from the discussion of biopower (his species chauvinism, as Haraway calls it), Nealon argues that in Foucault's work the animal is not excluded or forgotten with the birth of biopower, but quite the opposite: "animality is the dominant apparatus for investigating what life is and what life does" (Nealon 2016, 8). Nealon's immediate goal is to account for the rise of the importance of "the animal question" in contemporary scholarship, and he sees Foucault's biopolitics as providing a possible answer, if probed deeply enough. In fact, through a close reading of *The Order of Things* and *History of Madness*, Nealon argues that for Foucault in the larger biopolitical framework of modernity "animals function less as our ethical 'absolute other' than as our hidden, better self: our unconscious drives are animal in nature, and it is those unbridled desires that make us most truly who we are" (11). Yet, should one apply this framework not just to the emergence of animal studies (as something of a necessary consequence of affirmative biopolitics) but also to changes in dog training, pieces begin to fall into place. It is the emphasis on the nature of animality that proves crucial. While within radical behaviorism, training was about shaping the animal's desires in a way that makes them coincide with those of the human, the added element in the most recent methodologies is emphasis on the process of identifying, acknowledging and appreciating "animal desires." While the revaluation of animality was visible already in Nietzsche, and – as I argue in chapter 3 – was enacted in dog training practices at the opening of the twentieth century, what happens now is different because "animality" is differently defined. The category no longer hinges on Nietzsche's fascination with wildness, strength and violence but refers to what is perceived as an animal's absolute honesty, truthfulness and the resulting enjoyment of life. A dog's life is the kind of life modelled by Diogenes the Cynic; devoid of superficial needs and built on the immediate satisfaction of natural ones: the need to run, move, socialize. Of course, the explicit scientific grounding of this turn in dog training is not Diogenes but contemporary ethology, which has, however, changed significantly in the past two decades. The belief in the dog's natural proclivity for aggression, which fueled dominance-based training methods, becomes replaced with the view that dogs are peace-loving and opportunistic creatures.

Consciously refraining from applying Deleuze and Guattari's "becoming animal," I use the motto "be more dog" to put a name to this "animal turn" within dog training. This preoccupation with identifying and fulfilling a dog's needs coupled with the human's desire to experience the world from the dog's point of view – as opposed to the pastoral inculcation of properly civilized desires in the dog – seems to be the guiding principle in most progressive dog training methods. These changes also include a very vocal denouncement of the type of total control of a dog's life associated with radical behaviorist training. Critical of the popular behaviorist idea that dogs should work to earn even the most basic privileges in life, trainer Kathy Sdao defiantly titles her book and training philosophy *Plenty in Life is Free* (2012). This new order is no longer about teaching the dog self-control; that is teaching him to repress his instincts and urges. It is preoccupied with understanding where those instincts come from. Characteristically, this is accompanied by the emergence of numerous canine-related leisure and sports activities, which suddenly balloon from obedience (available since the 1930s) and agility (since the 1980s) to include: canine disc, dock diving, dummy trials, rally obedience, barn hunting, nosework, flyball, earth dog trials, speedway, frizgility, canine freestyle, dog parkour and many more. The proliferation of these activities again ties in biopower to neoliberalism: market demand shapes the supply of organized human-canine activities available to those who are able to pay for engaging in them. However, those who do become involved explicitly state that they do so with their dog's best interests in mind. The empirical research carried out on humans' motivation for participating in these events reveals that humans take the dog's personal preferences into account while choosing the training activity they will both engage in (Hultstein 2014; Farrell et al. 2015). While the dog engages in an activity he loves (dogs with a passion for digging holes in the backyard will excel at earth dog events, those with an inclination for chasing moving objects will excel at lure coursing), the human is often forced to step outside of his comfort zone and follow the dog's passions. As the examples analyzed in the last chapter – Cat Warren's memoir on training her cadaver detection dog Solo and Silvia Trkman's materials on agility – suggest, training becomes an exhilarating experience that allows a human temporary access to the canine's mind and even a vicarious experience of canine physicality. A new being, a new entity is born in the human-canine performance. No longer is the learning process unidirectional; the human changes as well. As Haraway writes of her own experience of agility, "both players make each other up in the flesh. Their principal task is to learn to be in the same game, to learn to see each other, to move as someone new of whom neither can be alone" (2008, 175–76). It is the openness to change – evident in Haraway's quote – that best summarizes the trend in emerging dog training methods today.

CHAPTER 1

The Gentle Way in Punishment: Transcending Animality/Performing Animality in Early US Pet Dog Training Manuals, 1850–1900

> We humans are consciously anxious about our status and so we do something utterly paradoxical to address this. We construct animals as beings like us in order to show how powerful we are in our control over them, and simultaneously we make it appear that our power is natural in that it is given by animals that we have also constructed as instinctive and not rational.
>
> ERICA FUDGE, *PETS*

•••

> In every home where there are children and a pet dog, some one is sure to be very ambitious to give that dog special training. The older brother or sister will patiently go over and over daily instructions and search for new tricks to "show off" with the pet; while the little people will be equally interested and delighted with the "doggie" that will submit to being harnessed to the little cart and give baby a fine ride out-of-doors. (….) The doggies seem so much like real people when they are dressed up. And how do we know that they do not think and dream and plan their work and their play, and even talk to each other in dog language? Did you ever see a dog laugh? Just watch the home pet the next time he is romping and playing and see how his eyes shine and smile, even if he does not laugh out loud.
>
> *A VISIT TO THE CIRCUS, WONDERFUL PERFORMING DOGS, THE RIVAL MUSICIANS AND OTHER STORIES*

∴

Reading through American nineteenth-century dog training manuals, it is easy to observe how entangled they are in debates on species difference, nature,

 | DOI:10.1163/9789004380295_003

race and heredity. These themes are discussed either overtly in sections describing the purpose of animal training or can be discerned through an analysis of the behaviors taught to animals, the methods employed in the process of training and the material circumstances connected to staged displays of trained animals' skills; which after Michael Peterson's notion of "the animal apparatus" (2007) I will refer to as the training apparatus. The discussion can be read as revealing a certain tension in the conceptual framework of these texts; a tension between training as a vehicle for the transcendence of animality and training as the performance of animality. On the one hand, training – specifically, state of being trained, the result of training – offered the animal a possibility of transcending his condition of animality, in particular understood as rising above purely instinctive behavior; learning to control "animalistic" impulses. Additionally, the almost human-like feats of "sagacious" dogs, particularly in acts that conspicuously anthropomorphized them, seemed to suggest the possibility of blurring the human-animal boundary. On the other hand, these animal acts operated through a literal performance of animality, which utilized and strengthened the spectator's preconceived notions of what it means to be human and what it means to be animal. Nineteenth-century training texts hint at the possibility of abolishing the categorical boundary between humans and non-humans as based on humans' capacity to reason and then re-establish that very boundary through what Birke, Bryld and Lykke (2004) refer to as the discursive process of "animaling," that is the cultural production of the human/animal divide (Birke, Bryld and Lykke 2004, 168). This chapter traces the tension within these manuals and contextualizes the emergence of dog training literature in the second half of the nineteenth century among debates on race, nature and heredity. To quote Kathleen Kete's *The Beast in the Boudoir*, a book on pet keeping in nineteenth century Paris, such books were far from "a prosaic tool for household husbandry, dog-care books functioned as maps for the imagination, manuals of world-making" (Kete 1994, 81). In the case of the United States, a reading of animal training as inextricably entangled in the understanding of human race seems very timely, as increasingly more scholars suggest that species functions as a framework for rearticulating fears and anxieties related to race (Wolfe 2003a, Glenney Boggs 2013, Kim 2015). I argue that in the second half of the nineteenth century, the very white and bourgeois pastime of teaching tricks to pet dogs ultimately serves the purpose of rearticulating not only species difference but also – through a sort of reverse re-enactment of blackface minstrelsy, which portrayed dogs as almost human and yet absolutely not human – the superiority of whites.

In addition to the Foucauldian framework presented in the introduction, which works as a linchpin for this book, the conceptual framework of this chapter is indebted to the notion of performativity of species, explored by Lynda Birke, Mette Bryld and Nina Lykke in the article "Animal Performances" (2004) and to Jane Desmond's analysis of marine mammal performances in *Staging Tourism: Bodies on Display from Waikiki to Seaworld* (1999). Birke, Bryld and Lykke draw parallels between discourses of gender/sexuality and animality and suggest that the notion of performativity, as developed by Judith Butler (1990, 1993) and later Karen Barad (2003) in relation to gender, can also be applied to an analysis of species. In short, Birke, Bryld and Lykke point to species as performative and, as an example, discuss the "laboratory rat" as always constituted performatively through interactions with the material context (the laboratory, the scientist) but also through biopolitical breeding practices: the rat's body is shaped as a result of breeding selection to fit the laboratory equipment; more docile and easier to handle rats are selectively reproduced, etc. While their analysis may not seem groundbreaking in that the entanglement of animality with race, gender and sexuality as well as notions of the cultural construction of species were explored already in the 1980s by Donna Haraway – a fact which Birke, Bryld and Lykke duly acknowledge – what makes their article a potent source of inspiration for this work is their emphasis on "the concept of 'performativity' [being] useful for analysing co- or intra-actions of human and non-human actors" (Birke, Bryld and Lykke 2004, 168). The trained animal act is always a performance of animality in relation to humanity. Within the framework of training discourses, "dogness" is performative; the dog becomes a dog through training but his animality is always relational, constituted in reference to the human partner and shaped by other discourses over which the individual dog and the individual human have no control. Thinking through a similar idea, Michael Peterson coins the notion of the "animal apparatus," to provide a framework for taking into account, in the thematic analysis of animal performances, aspects related to the production of such performances. In a theatrical performance involving animals, the animal apparatus pertains to both props used to teach and later elicit the performance but also elements of the stage sets that inevitably aid in the production of meaning. Peterson writes "collars, reins, bits, whips, food, treadmills are part of this apparatus, but so are lights, wings, and even the very concept of on- and offstage space." (Peterson 2007, 34). Training manuals are, in this sense not only part of the training apparatus, but also a record of its other constitutive elements.

Meanwhile, in *Staging Tourism* Jane Desmond looks in detail at contemporary marine mammal acts at Sea World and reads them as a performance of "nature." While the Sea World performances reflect a different and updated vision of nature than the one in nineteenth-century dog acts – the difference being, in brief, that the natural state in the late twentieth century is one of man's harmony with "nature" (whales), while the nineteenth-century version is one of man's natural dominion over "nature" (dogs) – what makes Desmond's analysis relevant for the topic of this chapter is her description of the process of training as the cultivation and disciplining of the "natural" in the process of training. Behaviors identified as best reflecting the "natural" relationship between man and animal and the "natural" features of the particular species, the essence of dogness (or whaleness), so to speak, are selected for in the process of training, to create a performance that simultaneously enacts species difference and suggests the possibility of obliterating it because the animals are revealed as amazingly smart, "almost human" (Desmond 1999, 237–38). In the case of dog acts, what is staged as "natural" is the animal's willingness to submit and please the human, though a close reading of the training instructions reveals this eagerness to be an effect carefully constructed through specific training technologies and selection of the trainee. Desmond also traces how acts based on anthropomorphism and mimicry of humans were gradually being dropped as the framing of the marine mammal shows shifted from entertainment towards a more educational tone (239–40). In the period under consideration here, acts based on mimicry and very literal anthropomorphism most certainly stole the show, and Desmond's observation that their power lies in revealing the "dissimilarity between the animal and ourselves," in a certain "as if-ness," (240) serves as a guiding principle of this analysis. A similar observation about the effects of animal performances has been made by Brett Mizelle in relation to late eighteenth-century animal acts in the US. Writing about primates, Mizelle notes: "[a]nimal exhibitions also served as a locus for the (re)conceptualization of ideas about race, gender, and hierarchy, particularly when shows of performing animals or of primates problematized the boundaries between human and animal" (Mizelle 2005, 221).

1 Dog Training in the Nineteenth Century

Mizelle's historical scholarship has focused on what may be termed issues of spectatorship; on the reception of the already fully trained animal; on the performance itself as a locus of social anxieties related to race, gender and species and on the growing ethical concern of the spectator (not the trainer) in light of the rise of what Katherine Grier terms "the ethics of kindness" to animals in the nineteenth century (Grier 2006, 128–82). What this chapter attempts to do is shift the focus of academic investigation onto the process of training itself and the human-animal relationship in the process of training. However, the trainer-trainee relationship cannot be discussed without taking into consideration the purpose of training. The final goal of the entire process is not just somewhere in the background but clearly shapes the training relationship. For example, in the first half of the twentieth century, at the time of the strengthening of the disciplinary framework of dog training, the seriousness of the undertaking of training was used to justify harsh training methods in the training of police and military dogs (cf. Konrad Most and his American school of followers). While the training relationship can (and in some contexts does) exist outside of relations of spectatorship, the emergence of the very texts under analysis here, late nineteenth-century pet dog training manuals, is closely related to issues of spectatorship; specifically, to the popularity of performances by trained dogs.[1]

In *Pets in America*, cultural historian Katherine Grier observes that animal training handbooks aimed at the owner of the pet dog, as well as general pet care books that often include chapters on training, begin to be published in the second half of the nineteenth century as the market of for popular household handbooks emerges in the US. Pet training manuals are predated by the publication of instructions (in the form of complete handbooks or sections in books with a broader theme) geared toward the owners of working dogs (Grier 2006, 74). These are predominantly related to hunting, which reflects the class-based distinctions between working dog owners; books on sheepherding from that period are extremely rare because the class status of hunters was much higher than that of shepherds and, consequently, facilitated their access to the publishing market.[2] The training of hunting dogs is often referred to as

1 Though also, it must be added, a significant role can be attributed to the emergence of a publishing market for popular handbooks. Some of the animal training handbooks were published as parts of series of general advice publications. John Biggle, the author of *Biggle's Pet Book* (1900), was a well-known writer of numerous advice books for farmers (Biggle Farm Book series) and Jesse Haney's *Haney's Art of Training Animals* (1869) was reprinted as part of Fm. Lupton's *The People's Handbook Series*. Francis Butler, the author of *Breeding, Training, Management, Diseases of Dogs* (1857) was also the author of popular foreign language textbooks and books of poetry. He did, however, work as a veterinarian.

2 The only nineteenth-century US handbook devoted solely to the training of sheepdogs that I have managed to access is W.A. Wickham's *Practical Training of the Shepherd Dog* (1891). Many of the handbooks for pet dog training contained anecdotes or second-hand information related to sheepdogs. For example, John Biggle admits to never having trained a sheepdog himself, yet he includes a chapter on the training of herding dogs as "some helpful suggestions gleaned from various sources and pronounced correct by those who have had practical experience in the matter" (41). Similarly, C.H.C. Williams, also admits to not knowing much about the training of sheepdogs but offers to "give you the knowledge of the above as we obtained it from a celebrated dog trainer" (181). Henry Stephens Randall, Thomas Jefferson's biographer, was also known for a book titled *The Practical Shepherd* (1863), which focused on sheep management but also included descriptions of different types of sheepdogs.

breaking and the more pet-oriented manuals which will be analyzed here repeatedly admonish the amateur reader to leave the training of hunting dogs to the "professional breakers." In the case of hunting dogs, the trained animal's performance in the field, while not entirely hidden from public view and subject to some form of evaluation, is not viewed explicitly through the perspective of providing entertainment for the onlooker but through the optics of the dog's usefulness for the human, even if aesthetics does play a role in the evaluation of use value.[3] The aim of the training books that emerge since the 1850s is to convey to the growing bourgeois pet-owning population the methods that could be applied to teach animals the behaviors seen in the circus and in performing animal acts. Some of these books are not species-specific (Haney's *The Art of Training Animals*) and provide instructions for the training of both wild and domestic animals, though with an emphasis on domestic animals; some combine information on the training of dogs with that on the training of horses (Williams's *New System of Training*) and some are general pet care books that include information on training (*Biggle's Pet Book*), but there also emerges a number of publications devoted solely to the pet dog, such as *House and Pet Dogs: Their Selection, Care and Training* (1890). The training section in the last book is a reprint of the section titled "Ladies' Pet Dogs" from S.T. Hammond's book on the training of hunting dogs. A number of articles on dog training were also published in the period under discussion in major dailies and magazines.

The fact that handbooks providing instruction in the training of animals for any willing reader appeared only in the 1850s, does not mean that animal acts were unheard of earlier or that pet dogs were not taught the behaviors described in the handbooks. Brett Mizelle recounts American trained dog acts from the 1790s (Mizelle 2005, 221). Michael Dobson, in a study of the use of animals in theatre and fairground performances in Britain, traces the first appearance of trained canine actors to the year 1762, a date that marks the staging of *Two Gentleman of Verona*, where the dog playing Crab was, for the first time, not brought onto the stage on a short lead but trained to perform his role. Dobson notes that by the late 1770s dogs had become frequent actors on stage but that their appearances were much more common in comedies. The association of canine actors with comedy and, increasingly, lowbrow entertainment became so strong that in England managers of serious theaters banned the use of dogs (Dobson 2000, 120). Mizelle, writing about the American context

3 For an analysis of the contemporary foxhunt as a performance, see Garry Marvin's extensive work on foxhunting in Great Britain, especially "A Passionate Pursuit: Foxhunting as Performance" (2003) and "Natural Instincts, Cultural Passions: Transformation and Performances in Foxhunting" (2000).

in the early nineteenth century also posits that "animal acts supplied tawdry, lowbrow entertainment" (Mizelle 2005, 220). While the status of animal performances does rise somewhat as the nineteenth century progresses, alongside the great success of P.T. Barnum, trained dog acts are never seen as having much educational value, as opposed to exhibitions of exotic animals. In fact, one of the reasons why the contemporary colloquial expression "dog and pony show" connotes disdain is because actual dog and pony shows emerged in the 1880s as something of a less prestigious (and cheaper) alternative to the "real" circus.[4] Such shows were smaller, focused on exhibiting domestic animals and attracted a less respectable audience. The fact that Henry B. Gentry, the owner of the largest and most famous show, advertised his business as "the only moral exhibition in the world under canvas. An educational festival patronized and endorsed by the elite of the land" (M.A. Smith 1947, 22) actually reflected the anxieties that surfaced around the propriety and respectability of such endeavors.

The stock figure of the dog trainer, as appearing in magazine illustrations and even on books covers, was also firmly associated with teaching animals to perform circus stunts. In the period 1860–1900 (and even beyond the end date) it was not common for trainers to offer their services to the public and, more generally, to instruct pet dogs and their owners together. Sometimes, though rarely, pet dogs of the wealthy elites would be sent off for training, which they received while being boarded at the trainer's kennel. In 1871, *The New York Times* published a rather extensive interview with a dog trainer, titled "Training Dogs: A Morning Chat with a Professional Dog Teacher." The claim that the training of dogs as a profession outside of the context of the circus was very rare can be supported through the author's surprise upon seeing the sign "Dogs Educated" affixed to a building. As curiosity tempts the journalist to interview the owner of the business, he begins with the question: "You do not mean you make a regular profession of it, do you?" while the unnamed trainer answers "Oh yes, certainly – though there ain't more than three in the whole country (...) most of them combines exhibitions with their animals and has a line of fancy dogs for sale." The contents of the article also explain what professional dog training entailed at the time. It appears that the trainer both sells his own trained dogs ("the finished product") and offers to take dogs in for training, although that line of activity is less popular because, as the trainer informs, "Americans has to be educated up to my art." When the trainer is asked for a

4 According to Janet M. Davis, in the late nineteenth century prices of circus tickets ranged from twenty-five cents to 2$. Dog and pony shows charged only 10 cents (Davis 2002, 34).

demonstration, he lets out a group of dogs who display various circus tricks, though the comedy of the performance results also from their inability to perform the tricks perfectly: a fat spaniel topples to the ground instead of sitting on his haunches; a poodle takes him by the collar and leads him to a corner. The entire interview is related in a somewhat jocular tone and the trainer's colloquial vocabulary and non-standard grammar – transcribed, as it appears, verbatim – reveal that the journalist does not perceive dog training to be a legitimate pursuit for a respectable citizen.

While the training handbooks do not overtly engage issues of spectatorship, it is implied that the context does change everything: the training of dogs for personal entertainment is presented as an ethically legitimate and even educational pursuit. In an early British book on dogs by William Youatt, first published in London in 1845, later reprinted multiple times and published in America with a foreword aimed for the American reader in 1887, the author presents dog training as an educational pastime. Writing of a British "gentleman" who taught his dogs incredible tricks, Youatt emphasizes that "the performances were strictly private. The owner of the dogs was a gentleman of independent fortune, and instruction of his dogs had been taken up merely as a curious and amusing investigation" (Youatt 1871, 109–10). Francis Butler, the author of *The Breeding, Training, Management and Diseases of Dogs* (a volume of 430 pages), writes: "I have trained many dogs for my own private use" (F. Butler 1887, 91). Such defensive explanations, focusing on the private nature of the performances, stem from class-based anxiety associated with the low status of public performances involving dogs.

Without exception, the textbooks posit the training of dogs for private entertainment as a wholesome and educational activity, particularly for young people. An 1893 article on dog training from the *Los Angeles Times* encourages training as a free-time activity for the youth. The article's byline reads: "Any Boy or Girl Can Do It with the Home Dog" (Marsh 1893, 14). In 1882 *Harper's Young People* also runs an article geared toward an adolescent audience "Some Hints on Dog Teaching" by Edward L. Stevenson. The epigraph to this chapter, a passage from a children's book originally published in 1855, also posits that the teaching of tricks to the family dog is an appropriate and educational endeavor for older children (Figure 1). In a twist on the by now very familiar Victorian idea, which originated even earlier in the writings of John Locke, that caring for an animal helps develop responsibility and kindness in young people (Serpell 1999). In *Civilized Creatures* (2005), Jennifer Mason reads middle-class white America's affection for pets in the mid-nineteenth century as an element of sentimental education: kindness to pets is supposed to educate young people in compassion for other human beings; teaching young boys the true meaning of friendship and giving young girls an opportunity to practice maternal affection (Mason 2005, 13–16). Some of the authors insist that training a dog helps develop a young man's (particularly man's) intelligence and leadership qualities. John Biggle claims that "[n]o boy's life is quite rounded out if he has not had the companionship of a gentlemanly, intelligent, honest dog. In shaping the character of life of some interesting pet, the character of the young person is formed as well" (Biggle 1900, 8). Commenting on the classic construction of American uniqueness as tied to the notion of wildness, Jennifer Mason suggests that one of the reasons why pets had been excluded from playing a significant role in the critical narratives of what constitutes Americanness is because of their association with domesticity, and – by extension – with women. However, the training of pet dogs was not unequivocally marked as a feminine endeavor. None of the handbooks are targeted specifically at a female audience, though *House and Pet Dogs* reprints the chapter "Ladies' Pet Dogs" from S.T Hammond's book for hunters. The women are encouraged to be firm and patient as "there is little fear that a lady will be unnecessarily severe with her pet and so it is not worth while to recommend her to spare the rod" (*House and Pet Dogs* 1890, 53–54).

The circus dog trainer is targeted for criticism in best-selling books published in the 1870s and 1880s, offering a glimpse of the insider's view into the circus life. These included Olive Logan's *The Mimic World and Public Exhibitions: Their History, Their Morals, and Effects* (1871) and John J. Jennings's *Theatrical and Circus Life; or, Secrets of the Stage, Green Room and Sawdust Arena* (1882). In order for dog training to begin operating within the realm of social power, it first had to be repackaged as a respectable activity that upholds dominant, middle-class, bourgeois values. After all, only if the trainer's own intentions were perceived as noble, could training serve the function of ennobling its canine subjects. Disassociating dog training from circus acts, promoting it as an activity for young people and ladies, as well as emphasizing its place in the construction of an ethics of kindness to all creation became a strategy for countering the image of the often cruel and certainly vulgar and greedy circus animal trainer, who becomes a stock character routinely depicted on the covers of American dime novels in the second half of the nineteenth century.[5] Still, the tension surrounding the respectability of the pursuit – and the trainer's noble intentions – does not disappear completely.

---

5 Jesse Haney's book *The Art of Training Animals* (1869) is an exception in that it does not disassociate the training of pet dogs from the circus life. On the contrary, the book is a guide for the training of both wild and domestic animals.

2 Canine Sagacity

Notwithstanding the positive effects of training on a young person's character, within the regime of social power, training is seen as primarily beneficial for the dog. As Katherine Grier writes, with reference to the second half of the nineteenth century, training is inevitably described through the metaphor of civilizing the dog, "'raising' an undeveloped mind into a state of concord with other, cultivated members of the community" (Grier 2006, 74). It is training that allows the animal to fulfill what is viewed as his hidden potential and to attain a state of almost human intelligence. A crucial word used in nineteenth-century discussions of animal intelligence is sagacity. Dogs are invariably seen as particularly sagacious, which already points to the fact that an animal's sagacity is strangely tied to his relationship to the human. In fact, Harriet Ritvo writes of sagacity as reinforcing human dominion over animals: "[i]t could be defined so that the animals that exemplified obedient subordination had the largest measure." (Ritvo 1987, 38). In an article on "sagacity" in nineteenth-century Britain, Rob Boddice argues that the concept was used in a broader sense than Ritvo suggests: not only to signify human dominion (animal sagacity versus human intelligence) but also to "establish human-animal contiguity through a form of 'natural' ingenuity (distinct from instinct), or [that] it could even mean 'reason' " (Boddice 2012, 70). In other words, sagacity was a catch-all concept that could be stretched and contracted to reflect the writer's personal viewpoint. Still, what remains undisputed about canine sagacity is that it could be increased as a result training, "persistent contact with man, coupled with breeding and training" (Romanes 1882, 437). Consequently, the training relationship, while emphasizing human dominion over nature is based on human benevolence and good will: the human does the dog a favor by increasing the animal's natural sagacity. In fact, some dog trainers posit themselves as the saviors of dogs who would otherwise perish without proper instruction. Francis Butler writes: "My object is to prove that it requires but a small amount either of time or labor, to rescue even the meanest mongrel from a life of degradation and misery and place him in a position, where he may be respected for his virtues and appreciated for his worth" (F. Butler 1877, 88). Training is routinely presented as improving nature. Jesse Haney wrote in 1869, in a handbook that received considerable attention and was later (1894) reprinted in a popular dime format, that it is "under the tuition of man" that animals can "be developed to an extent to which they would never attain in a state of nature" (Haney 1894, 1).

Unsurprisingly, this rhetoric bears and eerie resemblance to arguments justifying the spread of British imperialism; and also, to America's colonialist project; famously described by Rudyard Kipling as "The White Man's Burden." This rhetoric of saving the "brutes" from their own "wildness" was a staple element of the rhetoric of Manifest Destiny, and it also shaped the understanding of racial relations between whites and blacks. In the mid-nineteenth century, the rhetoric of civilization was so pervasive that is was used not only by proponents of the institution of slavery, but also by abolitionists: a famous example being the wild, animalistic, "heathenish" Topsy in Harriet Beecher Stowe's *Uncle Tom's Cabin*. Topsy's salvation, synonymous with a literal conversion to Christianity, comes when she accepts a lock of hair from the dying Little Eva. In *Sensational Designs* (1985), an early feminist attempt at salvaging sentimental literature, Jane Tompkins emphasizes Stowe's skillful use of sentimental rhetoric, a convention her female readers would have been familiar with through their experience with women's domestic novels. Tompkins sees Eva's deathbed scene as a brilliant reduction of American culture's favorite story of salvation through love (Tompkins 1985, 125). Of course, as later scholars noted with reference to *Uncle Tom's Cabin*, the gesture of love is still steeped in a certain symbolic violence: once Topsy allows herself to be affected by Eva's love, she must renounce her "black ways." Both civilization and reason come only through interactions with white culture. It is obvious that within the framework of *Uncle Tom's Cabin* that the lighter the character's shade of skin (the closer they are to whiteness), the greater their intelligence and the higher their status among fellow slaves. It is no coincidence that Topsy was "the blackest of her race," just as it is no coincidence that Butler wrote of "the meanest mongrel": the hierarchy of blood that was an inseparable element of the colonial enterprise exceeded species boundaries.

To return to dog training manuals, it is worth noting that the very act of being selected as the subject of training supposedly raises the chosen animal among other fellow dogs. The life of an untrained dog is likened to a state of degradation and misery, mostly because it is seen as one ruled by instincts and passions, which are the obvious mark of animality. Instinct is often opposed to reason and training is seen as enabling the animal's faculty for reason, thus – though only to a certain extent – of transcending the state of one's animality. Haney writes about the distinction between instinct and reason as a divide that training aims to overcome by providing animals with tools that allow them to develop reasoning skills:

> Dr. Kemp[6] thus concisely and clearly states the difference between instinct and reasons: "In the former there is an irresistible impulse to go

6 Most likely Johannes Van Der Kemp, a Dutch missionary to South Africa and a racial reformer, 1747–1811.

> through a certain series of motions after a certain fashion, without knowing why they are performed, or what their result will be. In the latter the actions depend upon previous mental judgments, are performed or not at will, and the end of them is anticipated and clearly defined" (. ...) Under the tuition of man the reasoning powers are undoubtedly developed to an extent to which they would never attain in a state of nature, and by judicious and persistent teaching numerous animals have been educated to an almost startling degree. (Haney 1894, 1)

Here, instinct – also associated with impulsivity – is clearly equaled with lack of reflection, lack of thought, and is certainly not a desirable trait in either human or animal. Seizing control of an instinctive response becomes a mark of reason. George Youatt, already in 1845, writes of the disservice done to dogs whose natural instincts are not curbed. For example, Youatt sees the killing of a sheep as "nothing more than a perfectly natural impulse unrestrained" (Youatt 1871, 25). These ideas about impulsiveness opposed to self-control reflect broadly understood Victorian values, among which self-restraint had an important place. It is up to the human to take the leading role and help the dog in attaining a higher state of being. It is also not coincidental that within the discourse of scientific racism the supposed faults of African Americans – the ones that animalized them – were their supposed instinctiveness and lack of self-control, as explored in more detail in the next chapter. Submitting to human will results not only in the development of reasoning skills but also in the canine's happiness. Butler ponders: "Is it not worth the while to rise a quarter of an hour earlier (if necessary), to educate a faithful friend, who will be rendered the happier, by devoting his whole life to your useful service?" (129). As Harriet Ritvo memorably wrote, the Victorian love affair with dogs resulted from the perception of canines (next to horses) as the most malleable, the most plastic, the most receptive to human intervention and – at least outwardly – happy with the results of such intervention (Ritvo 1987, 115). The belief that dogs benefit from instruction and are eager to please their masters when they have been properly, that is without the use of excessive force, taught how to serve them was predominant also in in the United States.

At the same time, reason was supposed to result not only from suppressing instinctive behavior (as Butler's example with the chasing of sheep may suggest), but also from repetition of instinctive behaviors in controlled circumstances: "a purely instinctive action never changes except under the influence of reason" (Haney 1894, 1). Haney gives the example of hunting and coursing dogs who learn through experience; for example, as a result of performing the instinctive chasing behavior multiple times they learn to predict the hunted prey's behavior which, in turn, alters the route they take on the chase. Yet nowhere does Haney explain how this principle applies to the behaviors he proposes to teach the pet dog. One may infer that Haney could believe that the instinctive response (following a treat or shrinking from a whip) could become an expression of reason once the props eliciting the response (treat or whip) have been removed and yet the behaviors persisted.

The previously cited lengthy quotation on reason from Jesse Haney's book is not the only one which attributes the possibility of reasoning to animals in dog training literature of the period: notwithstanding the muddled and unclear definitions of what constitutes reason. George Youatt describes the dog's "natural endowments" as "not merely combining memory with reflection – but possessing qualities of the mind that stagger us in contemplation of them" (Youatt 1871, 105) and all the other authors sing the praises of canine intelligence. One would be tempted to account for such views through the influence of Darwin's *Descent of Man*, which not only states directly that "the difference in mind between man and the higher animals, great as it is, certainly is one of degree and not of kind" (Darwin 1981 [1871], 105) but also goes on to give dogs as an example of the animal which has managed to progress to a more rational state of existence than his predecessor the wolf, as a result of interactions with man. However, Darwin's book came out two years after Haney's book was initially published. And even though some of these ideas can be traced back to the earlier *On the Origin of Species* (1859), the view of dog training as a civilizing endeavor that targets a subject hypothetically capable of some reasoning powers cannot be a strictly Darwinian influence but rather a broader Victorian paradigm, as proved by the fact that George Youatt's 1845 book paints a picture of the dog very similar to that later presented by Darwin.

3 The Gentle Way in Punishment

In *Civilized Creatures* Jennifer Mason turns to Foucault's discipline to describe the changes in the methods for horse-breaking in the second half of the nineteenth century; she reads the books as progressing from reveling in an almost sadistic violence to a use of force that was more subsumed and rooted in utility: use only as much force as necessary in order to obtain the desired effects. Writing about American horse breaking manuals, Mason argues that: "The modern system of horse management is a nearly perfect manifestation of disciplinary control. Trained in this system, equine bodies are docile bodies" (Mason 2005, 35). This does not seem to be the case with dog training manuals, in which is it hard to talk about the progressive gentling of methods in the second half of the nineteenth century: it will become clear that declarations of love and kindness clash with the brutality of the actual methods. Mason's book provides a rather generous reading of American sentimental culture, in which the kind treatment of animals becomes both a lesson in humanity and a marker of middle-class status. However, studying dog training manuals from the same period reveals a tension at the center of the sentimental project: the rhetoric of "salvation through love" continually clashes with "regeneration through violence," to allude to the title of Richard Slotkin's classic reading of American culture. Violence is necessary in order to achieve the effect of "salvation through love." This violence appears to be a structural element of the rhetoric of civilizational progress and if there is anything that a reading of dog training literature can contribute to our understanding of American culture in the nineteenth century, it is how literal this violence is: how much force was applied in order to create the "naturalness" of canine submission. Going back to the epigraph to this chapter, in which happy doggies dance on their hind legs and pull carts, it is almost certain that the textbook consulted by the young trainer would have advocated judicious use of the whip alongside declarations of dogs' natural sagacity and love for humans.

In the mid-nineteenth century it does not yet seem valid to speak of Foucauldian discipline in relation to dog training: the methods are too messy; the elegance of discipline can hardly be discerned. As I have argued in the "Introduction," in the case of dog training, the second half of the nineteenth century is representative of the changes described by Foucault in "The Gentle Way in Punishment" chapter of *Discipline and Punish*: there appears a certain reflection on the impropriety of excessive force, which obviously clashes with the proclaimed values of the period, but no force-free system appears. In fact, I argue that it cannot appear, simply because violence is a structural element of the rhetoric of civilizational progress, or, as some would argue, of humanism. Additionally, gentleness can be a notoriously subjective concept to define, as will become clear in the chapters to come, when I speak of the controversy surrounding contemporary "positive training."

However, the rhetoric of the gentling of methods can certainly be discerned on the declarative level, where gentleness is tied to civilizational progress. It was not just methods for the training of dogs that were said to be evolving toward gentleness; the methods for the training of wild animals were allegedly changing as well. Reading circus trainer Carl Hagenbeck's well-known memoir *Beasts and Men* (1909), Peta Tait interprets changes in the training of wild circus animals, that is the introduction of gentler methods and food rewards as well as a more general paradigm change to training viewed as education, through the shift in vocabulary from taming to training (Tait 2012). Elsewhere,

Susan Nance reads the change in the vocabulary of circus elephant trainers – from training to teaching – as more of a public relations strategy than a marker of actual change (Nance 2013, 72). Tait reads "gentling," as Hagenbeck described his methods, in connection with the emergence of the vocabulary of education and the teacher-pupil relationship beginning to be applied to animals (Tait 2012, 15). This, in turn, she ties in with the emergence of popular Darwinism, according to which "the trained exotic animal was considered to demonstrate an evolutionary improvement, so animal training became associated with progress." Still, the animal act itself "reiterated a hierarchy in which humans could bestow evolutionary progress on a lower order of wild animals" (16). In other words, training begins to be marketed (at least by the trainers themselves) as enabling the animal to move up in the evolutionary paradigm, in a certain sense to transcend his status as an animal; yet the human's mastery of the training process and the human's role in the animal act restage a relationship of domination. It should be noted that Nigel Rothfels, in his account of the training of circus animals, questions whether Carl Hagenbeck is to be credited with the introduction of gentler methods (Rothfels 2002, 156–57).

In the world of dog training, the terms training and teaching had been used interchangeably for centuries. In fact, one of the first books about dogs in Britain, *De Canibus Britannicis* (1570), published in English as *Of English Dogges* in 1576, used the terms teaching and training, not breaking. Still, a decrease in the popularity of the term breaking in relation to dogs can be observed in the late nineteenth century. However, changes in the world of dog training were not a carbon copy of changes in the world of exotic animal training, primarily because the training of dogs was never seen through the optics of taming and, unlike the training of tigers or lions, was not described (until the early twentieth century and the emergence of Konrad Most's methods) as requiring skills such as courage and strength on the part of the trainer. As previously stated, the implied reader of much training literature was a young (implicitly white) middle-class male. Additionally, the training of dogs was clearly influenced by the assumption of the dog's "natural" submission: there should, after all, be no need to enforce a quality that was already present. Of course, the analysis of the training methods reveals that this "natural" submission was obtained through the use of the whip, but the declarative level is at odds with these practical instructions.

Furthermore, when one looks at the dog training books published between 1850 and 1900, there is no chronological progression in the gentling of training methods. Here, I will look at five books: Francis Butler's *Breeding, Training, Management of Dogs* (originally published in 1857), Haney's *The Art of Training Animals* (1869), Williams's *New System of Handling* (1878), Biggle's *The Biggle*

Pet Book (1900), *House and Pet Dogs* by Forest and Stream Publishing Company (1890) and the 1882 article from *Harper's Young People*. These texts present a range of methods for teaching a surprisingly stable set of exercises, all of which rely on canine imitation of human actions, that is on exercises that literally anthropomorphize dogs, who are taught to exhibit behaviors as if – to go back to Jane Desmond's notion of "as-ifness" – they were human. The most popular exercises include: shaking hands, sitting up, walking on hind legs, playing dead, jumping through a hoop or over a stick, speaking on command, bringing various objects, taking a bow, walking on front legs, creeping (crawling), climbing a ladder, balancing a lump of sugar or treat on the dog's nose. Some also include coming when called, sitting and lying down. All of these texts also contain an almost inordinate number of anecdotes about smart dogs who could play cards, dominoes, spell words, go shopping on their own, etc., without providing specific instructions on how to teach one's own dog such complicated tricks. It is interesting that at this time the anecdotes do not celebrate canine feats of improbable heroism but recount hard-to-believe performances of activities that seem to require human reasoning skills. Some handbooks also include relatively brief instructions related to the training of hunting dogs, herding dogs, watchdogs and fighting dogs (F. Butler) but all focus on teaching tricks to the household pet.

With regards to the methods used, all of the books – except Williams's *New System of Handling* – express an explicit preference for kindness. Haney writes, "When he [the animal] knows what you want him to do he will in almost all cases comply with your wishes promptly and cheerfully. For this reason, punishments seldom do any good" (Haney 1894, 2).[7] Butler encourages owners "to use as little force as possible" (F. Butler 1877, 88). John Biggle writes that "the way to a dog's heart is through his stomach" and "the hope of a reward is a far better educator than fear of punishment" (Biggle 1900, 45). In all cases, the use of brute force is presented as backwards, while techniques relying on rewards are marked as more progressive. Haney admonishes: "This education may be either in the primitive style illustrated by the administration of a vigorous kick (…) or the more elaborate training which culminates in the production of a canine prodigy" (Haney 1894, 20). Kinder treatment of the trained animal is univocally equated with progress, while force and violence stand in for primitive savagery. Butler claims that too much force will impregnate the dog with "a sneaking timidity," which is "degrading to the brute and disreputable to his

7 It should be noted that Susan Nance in *Entertaining Elephants* views Jesse Haney's book as purposely presenting highly sanitized methods of training because it offered a pro-circus argument in the debate on the training of circus animals.

master" (F. Butler 1877, 104). While this is a comment made in passing and not elaborated upon in the handbook, it may be read within the framework of kindness as progress, while brutality is equated with turning back the wheels of civilization; a framework which was also at the roots of turn of the century beliefs about (human) race and white racial superiority. The "white man's burden" of civilizing the savages extends across species barriers.

Yet, despite declarations of kindness, all four of the textbooks and the magazine article also describe the type of physical punishment to be administered upon the dog's misbehavior, and in all cases the primary means of administering punishment is the whip: an inevitable accessory of all dog trainers at the time. Some of the writers resort to the use of the whip for teaching even the simplest behaviors. For example, Williams uses the whip to teach the dog to respond to his name and come to his owner: "The proper way of training him is with a whip, a small riding whip. In giving him his first lesson, should he attempt to run from you (as he undoubtedly will give him a sudden jerk and say 'Come here'" (Williams 1878, 187). But even Haney (1869, 1894) and Biggle (1900) resort to using "a switch" in the teaching of more complex behaviors. Haney uses the switch to teach the dog to "stand on his head and to walk on his fore legs" (1894, 26), while Biggle advocates the use of a switch for teaching the dog to resist food and hold a treat on his nose. Biggle is also a precursor of the use of a rolled-up newspaper to punish the dog for various misdemeanors; a technique of punishment that becomes ubiquitous in the mid twentieth century. The *Harper's* article is a striking example of the ease of switching from food inducements to compulsion, an example all the more salient as it is addressed to children: "Take a dozen bits of crackers and (…) call him by his name. (…) If he refuses to come after you have called him (…) [g]o up to him and administer one single cut with your whip. It is well to use one whip throughout your dog's pupilage" (Stevenson 1882, 823).

In addition to the whip, all of the authors use food in training, though to different extents. Even Williams, whose "new system" is clearly the harshest of all of the training methods discussed, admits that the easiest way to teach a dog to "waltz," that is to walk on his hind legs, is to hold a piece of meat above his head, even though he is quick to point out that the dog should soon perform the exercise without a reward present. Haney mentions matter-of-factly that: "All trainers make use of various little tidbits as rewards for successful performance of tricks" (1894, 2). The other writers use food as a lure for many of the behaviors (sitting up, walking on hind legs, crawling, jumping through a hoop). The lure is then faded out as the dog becomes familiar with the trick. The treat is later used as a reward and the frequency with which rewards are used depends on the quality of the dog's performances. Most of the authors suggest that rewarding a particularly good performance is a useful idea. However, as the dog advances in the process of training, he is expected to understand that obedience is a canine duty. For example, Butler warns that a dog may be spoiled if too many people throw objects for him to fetch and he "considers it only a play-game and becomes remiss in his duty to his master" (89). The supposed gentleness of the methods – which is questionable bearing in mind the frequent and inevitable use of the whip – is part and parcel of the multi-faceted strategy all the books engage, that is of presenting training as a respectable pursuit for white middle-class American gentlemen and women, the bourgeois readers of the textbooks. It is clear that the real beneficiaries of the "gentle way in punishment" are not the animals but the humans who experience an increase in self-esteem as a result of using "enlightened" methods.

Still, what seems the most striking in the textbooks described above is not the uniformity of training methods, for this is not the case as the methods are quite varied, but the uniformity of exercises to be taught: tricks. Again, it is difficult to speak of discipline when utility is taken out of the equation. A docile body is, after all, a useful body. Tricks, however, are the epitome of uselessness; of entertainment for its own sake. While the pet is generally a "useless" creature, the pet manuals of the next generation (from the early twentieth century onwards) actually stray from the teaching of tricks specifically to mitigate this problem, in an attempt to re-embed training within the framework of utility. The word trick itself connotes deception and illusion, and the deception that seems to be in effect here has to do with the animal's imitative performance of humanity. This is the moment where the ethics of kindness tied to civilizational (or evolutionary) progress – so readily espoused, at least in writing, by the proponents of dog training – truly reveals itself to be a fluke, or at least a smokescreen. Despite the grand pronouncements of training being a tool for developing the dog's reason, what the textbooks propose are imperfect enactments of humanity; sending a message that while dogs may be sagacious, they will never be human. The learned behavior is always a trick, and one that humans immediately recognize as such, reacting to it with laughter. As previously mentioned, canine performances quickly became associated with lowbrow comedy. While the removal of the performance from the public eye may have lessened class-based anxieties associated with viewing animal acts, it did not alter the primary purpose of training. The purpose of circus acts was human amusement and the purpose of the teaching of tricks to dogs in private remained entertainment, despite declarations to the contrary and attempts at sanitizing the process of training through its embedding in the ethics of kindness to all creation.

This becomes evident when reading closely through the instructions for teaching some of the tricks. Firstly, the ease with which the methods for the

FIGURE 1. As this image from a children's book proves, teaching tricks in which dogs imitated human behaviors was promoted as a wholesome form of family entertainment. ILLUSTRATION FROM *A VISIT TO THE CIRCUS, WONDERFUL PERFORMING DOGS, THE RIVAL MUSICIAN AND OTHER STORIES*. 1899. PHILADELPHIA: FRANKLIN BOOK COMPANY. UNPAGINATED.

teaching of the more difficult tricks turn to the whip reveals the trumping of the animal's well-being with considerations of performance. Even though none of these handbooks explicitly speak of animal agency and training as communication or dialogue, they all declaratively recognize dogs as sentient and reasoning creatures. Yet, this view is completely absent from the specific instructions for teaching the tricks, where it is ease of eliciting the behavior using the simplest means possible that is the guiding principle. Secondly, the material reality of the performance, the training apparatus, again suggests the primacy of human entertainment. For Michael Petersen, the animal apparatus includes both the material objects used to create the performance and the material context of the performance itself. Francis Butler's description of the trick of sitting up in a chair is a case in point of how the training apparatus bears both on the training relationship and on the performance itself. Butler begins by teaching the dog the sit-up position: "to make a dog sit up, he must be placed in position and seriously threatened, should he dare disobey the order" and then adds: "He may then have a pipe placed in his mouth and a cap on his head; he may also be ornamented with a pair of spectacles, which together with a newspaper placed before him, will render him worthy of your attention" (104). What renders the dog worthy of human attention is the comic (and "cute") appeal of a dog dressed up in human clothing: the placing of the human paraphernalia on the dog is not even, strictly speaking, training and yet it is included in the instructions for teaching the trick precisely because it heightens the anthropomorphic and comic effect of a dog sitting up in a chair. Listing the distinctions between the disciplinary regime proper and social power, the transitional modality between sovereignty and discipline, Foucault mentions different primary practices of the two regimes: representation in social power and training in disciplinary power. The performance described above is symptomatic of this difference: what matters is not how the dog is trained but what his performance represents: the staging. Jeffrey Nealon talks of thematizing the shift from social power to disciplinary power as "the movement from representation to training" (Nealon 2008, 30). Here, we are deeply within social power. There is hardly any training to speak of. It is the staging of his body as human that renders the dog worthy of human attention.

Yet the purpose of the anthropomorphism, understood here as the dog's imitation of human form, is not to blur the human/animal divide, but to emphasize the abyss between the two species. While there has been a recent wave of scholarship on the positive possibilities of anthropomorphism as a concept that has the potential for improving the human-animal bond by making it more nuanced and attentive to the animal's needs (Mitchell 1997; Serpell 2005), this is clearly not the case here. Jane Desmond's reading of the anthropomorphic behaviors in marine mammal acts fits the nineteenth-century scenario perfectly:

> The more explicitly anthropomorphized behaviors (…) take their meaning from the humanness of the actions and the nonhumanness of the performers. In each case, bodily difference of the animal is the foundation through which the action gains its meaning. Whatever an animal

> does ultimately reaffirms our concept of it as an animal, given the resilience of the human-nonhuman divide. (Desmond 1999, 174)

These tricks are a performance of species difference, not a promise of transcendence. They are also, per the framework suggested by Birke, Bryld and Lykke, a performative enactment of dogness as perpetually entangled with and inseparable from humanity, but here obviously placed within a hierarchical structure of domination. Dogness becomes the desire to please the human, to submit to the human's wishes, but also to be like the human. In this performance dogness is tragically – or rather tragicomically – tainted with its own impossibility. Comically, because this "flaw" is the source of the spectator's laughter. Tragically, because no matter how hard the animal may try, he will never succeed. The attempt is doomed from the start.

4 Canine Minstrelsy

Even if we agree that coercion is a staple of all training methods – particularly if one defines coercion as broadly as Paul Patton does in his essay on the training of horses, that is as causing an animal to behave in ways in which it otherwise would not (Patton 2003) – in these textbooks there is an abyss between the level of declaration and the level of practice. And not just in declarations of gentleness and practices of cruelty, but on an even more fundamental level: a discrepancy between goals and effects. The theory of training seems to suggest a promise of the transcendence of animality, of the ennoblement of the canine through the development of his reasoning faculties. In practice, the tricks, and the material practices of training that go into creating them, degrade the dog by turning him into a flawed imitation of a human. Yet, one may ask whether this is simply not a contemporary (mis)reading of nineteenth-century practices? In other words, were the trainers aware that the selection of the behaviors (tricks), the methods used to teach them and their performances ("the training apparatus") did not add but rather took away from the perception of dogs as creatures capable of reason and possessed with an innate dignity? Maybe the authors really did perceive themselves as helping the animals develop their faculties?

It is impossible to answer this set of questions without returning, yet again, to the entanglement of nineteenth-century discourses of animal training with the discourses of race and nature. Some of the textbooks include seemingly odd and brief statements about the suitability of various types and breeds of dogs for trick training. Biggle closes his chapter on training with the following statement: "Hunting dogs should not be taught tricks. Collies are not good trick-dogs, they are too dignified" (Biggle 50). A similar view is briefly expressed by Williams: "A well-bred dog is hard to learn any tricks except those pertaining to his nature; while a mongrel cur is quite easy to learn any" (Williams 186). The authors of *House and Pet Dogs*, writing about Mastiffs, Newfoundlands and St. Bernards, explain that "there is a dignity about these large animals which makes the performance by them of mere tricks seem out of place" (62). Going farther back in time, even *Of English Dogges* presents crossbreds as best capable of learning tricks. After listing the various types of dogs and their uses, Caius finally lists mongrels, who, unlike the other dogs, have no proper higher purpose but who excel at tricks: " 'Curres of the mungrell and rascall sort' which are taught by 'their vagabundicall masters' to dance, to lie down at command, and, especially, to beg" (Caius 1912, 34–35; quoted in Dobson 2000, 118). The teaching of tricks, understood as behaviors that serve no purpose other than entertainment and that are based on imitation of human behaviors, is clearly – and not just in the nineteenth century, as the quotation from the sixteenth century treatise suggests – seen as demeaning and degrading to the dog's nature. As such, it should be the province of those who are already devoid of any dignity: mixed breed canines. The common and clearly long-standing association of tricks with mongrels suggests that the authors of the textbooks must have been aware of the discrepancy between their declarations of training as the development of reason and the reality of the performance, and this is why they discouraged trick-training of purebred dogs, who were seen as possessed with a greater natural dignity through their associations with human aristocracy, and who served as stand-ins for the owner's aspirations of social advancement (Ritvo 1987, 82–124). Yet, it also suggests more parallels between the discourses of human race and canine breed.

A lot of work has been done recently within animal studies pertaining to the messy (non)correspondence of race and species; research that can be summed up as complicating the parallel of "racism = speciesism" first articulated by Peter Singer, then developed in Marjorie Spiegel's *The Dreaded Comparison* (1988). Aware of how this rather simplistic analogy has spurred a series of much more elaborate investigations into the complexity of multiple intersections of discourses of race and species (cf. Glenney Boggs 2013, Kim 2015) I would still like to posit a certain similarity between the functions served by the instructions for the training of tricks in the training handbooks discussed here and the ideological function of blackface minstrelsy. This similarity rests on two nodes of intersection: the use of imitation as a strategy and the effects of the performance. Additionally, it is also not coincidental that the two types of performances arise in the US at more or less the same time. In her book about circus elephant performances in the US, Susan Nance sketches the connection between animal performances and blackface minstrelsy. Nance writes: "the practices of marketing and consuming nonhumans in these ways were as central to the development of American culture as blackface minstrelsy." Nance goes on to argue that: "[a]s the blackface form helped people to define race in the United States, Americans employed animal celebrity to complicate or to clearly mark out the imagined lines distinguishing humans from nonhumans" (Nance 2013, 4). Nance's observations seem even more suited to a discussion of canine performances than of elephant acts because mid-nineteenth-century dog training, as shown above, routinely enacted a flawed and comic impersonation of humanity; while wild animal performances did not. The roots of blackface tie the genre closely to the circus as well; as Eric Lott writes in his classic study of minstrelsy in America *Love and Theft* (1993), the first blackface performances actually took place between circus acts and "many minstrel performers began their career in the circus" (Lott 1993, 24–25).

While it is obvious that in canine performances it is the dog's impersonation of white humanity that generates the laughter and in blackface minstrelsy it is the white minstrel's impersonation of blackness that leads to the outbursts of merriment, what makes it possible to sketch a comparison of the two is the similarity of the effects of these two types of performances. While both show an artificial and exaggerated performance of, respectively, race and species, one which could potentially upset the social order through a depiction of racial/species transgression, both are inherently conservative in terms of their influence on the social order. Human minstrelsy "reveals" African-Americans to be inferior versions of whites: less civilized, less dignified and somewhat animalistic; canine minstrelsy seems to have had the same consequences for dogs. Instead of opening up the possibility of the collapse of the human-animal barrier, it reinstitutes the boundary with huge force. Both performances are controlled by those who consider themselves superior in the hierarchy of creation. In both, it is white middle-class humanity that serves as the norm which is aspired to by others. Both types of performances are, in short, re-inscriptions of domination.

Alongside pronouncements of the dog's loss of dignity, the textbooks do offer some vague explanations of why this is the case. These primarily go back to the Cartesian notion of animals as machines. *House and Pet Dogs*, for example, contends that tricks are "merely mechanical" (65) and as such cannot function as proof of reason but only of imitative skills, "cleverness," which is not equated with true intelligence. Additionally, while many of the books contend that curtailing instinctive behaviors is the very purpose of training and control of instinct is a mark of reason, a dog who acts against canine instincts, is seen as "unnatural." The subject of the training process is thus caught in a double bind: on the one hand, it is the dog's "nature" to be servile, but, on the other, being too servile, and participating willingly in the performance of tricks, causes him to be marked as "unnatural."

The manuals also reveal a curious tension in beliefs related to heredity and, specifically, to the inheritance of learned behaviors. In *The Platypus and the Mermaid* Harriet Ritvo explores the British Victorians' strange and highly sexist – from a contemporary perspective – presumptions about the breeding of animals: including the belief that the female's previous mate influences offspring in subsequent matings (telegony), the highly popular notion that the sire contributes more than the dam to the appearance and behavior of the offspring (Ritvo 1997, 104–113). Most importantly for this analysis, any type of miscegenation between species or between different breeds within a species was deemed an abomination. Ritvo recalls the words of an agricultural expert who "described crossing as 'a national evil and a sin against society,' especially if the cross-bred animals were themselves used for breeding" (105). At the same time both behavior (including learned behavior) and appearance (including altered appearance)[8] were seen as heritable. These beliefs, that is a proscription against miscegenation and a belief in the heritability of acquired behavior, created a tension that the authors of the training textbooks could not resolve. In the world of hunting dogs (and other working dogs), this tension was nonexistent, and the line of reasoning was clear: aptitude for hunting is passed on from generation to generation and can be enhanced through selective breeding to other dogs who exhibit the same behavioral traits and who, in the mid-nineteenth century, already happen to be described using the vocabulary of breed. Even S.T. Hammond, who based his training on what we would nowadays consider mostly behaviorist principles, reminds the hunting enthusiast to begin with a well-bred dog whose parents "are first class field performers" (Hammond 1882, 2). However, authors of trick dog manuals find themselves caught in something of a trap: on the one hand, beliefs in heredity are very strong; on the other, the breeding of "mongrels" is highly frowned upon while the teaching of tricks to purebred dogs is seen as taking away from their dignity. As a result, some persist in their beliefs about the heredity of acquired behavior, while others abandon them. Jesse Haney was a member of the former group and claimed that "[t]he descendants of an educated animal have, by inheritance, a greater aptitude for learning than others" (Haney 1894, 1) while Francis Butler snickered at a neighbor who believed that acquiring a puppy of

8 Butler writes about having raised "a litter or two of three legged puppies, from a slut who had unfortunately lost a limb" (14).

the same stock as a well-trained dog would automatically result in the puppy "educating himself, at a very early age" (F. Butler 1877, 88).

These mixed views result from something of a Gordian knot of intersecting beliefs about race, breed and heredity that cannot be untangled to the satisfaction of all those involved, but that show obvious parallels with nineteenth-century ideas about human race: on the one hand, the cultural proscription against miscegenation and, on the other, the emerging Darwinian concept of hybrid vigor. When Donna Haraway writes about the intersections of race-gender-species in "Teddy-Bear Patriarchy: Taxidermy in the Garden of Eden, New York City 1908–1936" she notes the paradoxes of the eugenicist and conservationist discourses emerging in the early 1900s: they attempt to incorporate the concept of scientific progress into the enterprise of perpetuating a timeless "nature," at the same time carefully constructing that "nature" to reflect dominant beliefs about race, class and gender. The trick training manuals under discussion here perform the same maneuver because they are invested in the same project: by virtue of their white middle-class positioning they perpetuate beliefs about the superiority of racial (breed) purity, which they also construct as timeless and "natural." Not surprisingly, the practical experiences of training professionals are at odds with their declarations. Not only does it turn out that performing does not ennoble the trainees; the supposed superior intelligence of purebreds is also questionable as many of the trainers admit that their best-trained dogs were mutts. Some trainers – as the previously quoted 1872 *New York Times* interview with a professional trainer reveals – breed their own lines of dogs and many, in practice, express a preference for poodles over mixed breeds. The unnamed New York trainer claims: "No dog is as good as a poodle. You would be surprised to know what a difference it makes when I train a puppy from intellectual stock." The poodle, even though a purebred, seems well suited to parlor tricks, possibly because his dignity is already seen as tainted through his association with women. The second half of the nineteenth century is a period when poodles definitely cease to be viewed as working retrievers and become firmly associated with circus performances and fancy lapdogs.[9] It seems tempting to draw the conclusion that the poodle's shift in status from a man's working dog to a lady's lapdog makes him lose some of the dignity of purebred dogs: the poodle begins to be associated with

9 Kathleen Kete describes a similar process taking place at exactly the same time in the poodle's homeland, France. The poodle begins to be disassociated from the brabet, which remains a hunting dog and a guide dog for the blind, while the poodle begins to enjoy "unrivaled popularity among Parisians by the end of the century" (Kete 1994, 87).

imitative behavior and "cleverness," which is deemed the opposite of true intelligence.[10] Meanwhile, many professional trainers prefer to choose their dogs among strays, a trend that continues well into the twentieth century. Still in 1928, Courtney Ryley Cooper in a book offering an insider's view on the circus life talks about how performing dogs are chosen from among the strays that follow the circus and from among the residents of animal pounds. Cooper also believes that while some types of dogs are better suited for certain types of acts, a pedigree hinders rather than helps in the learning of tricks. He closes his discussion of dogs in the circus with a curiously ambivalent statement: "around the circus you will find many a pedigreed dog of the various species but rarely a dog with a pedigree" (Cooper 1928, 124).

5 Conclusion

A close reading of dog training manuals makes it possible to analyze animal acts not just as performances, with a focus on the spectator's reception of the performance and its meaning-making within the broader context of discourses on class, race and species but to see how the performances are constructed: both in terms of the actual methods used to train the animals and of the beliefs grounding these methods, the ways of "world-making." In other words, it becomes possible to take into account the entire training apparatus and analyze how it performatively constitutes animality. These performances are most certainly not subversive reiterations of animality – to allude to Judith Butler's terminology for describing performances of gender – that is they strengthen rather than question prevailing beliefs about what it means to be a dog and what it means to interact with a dog. As something of a side effect, they also reveal certain tensions within the dominant discourses of the period: especially through their strong entanglement with discourses of race.

These include the implication of the ethics of kindness to all creation in a basically colonial project of educating the savages (both human and nonhuman) and the internal incoherence of that project as reliant, to various degrees, on violence. The textbooks may proclaim that gentleness is the preferred method of "educating the beast," yet when they reach the practicalities of the training process, they all turn to the whip. Another paradox is related to the

10 The association of the poodle with mechanical imitation and servility leads to further entanglements of race and breed. The African American writer Richard Wright in the essay "Blueprint for Negro Writing" (1937) postulates that the Negro writer is received by whites as "French poodles who do clever tricks."

understanding of the purpose of training. On the one hand, the process of training is supposed to elevate the canine's mind by increasing his capacities for reason and thus making the dog more like a human, within a Cartesian framework that clearly privileges reason. Yet what truly happens is that the dog is anthropomorphized not so much through the cultivation of his reasoning faculties, but as a result of a very literal imitation of human behavior; the performance turning into an act of "canine minstrelsy," clearly degrading for the dog. The dog is read as engaging in an "unnatural" act. Finally, the training apparatus points to contradictions in nineteenth-century understandings of heredity and to the implication of this notion in a basically racist (and clearly speciesist) biopolitical framework, where the cultivation of one's mind through training seemingly can, but on the other hand cannot, have bearing on the status of one's offspring in the hierarchy of creation.

I have argued for the use of the Foucauldian notion of social power as the framework for theorizing dog training in the second half of the nineteenth century. The use of the word social in Foucauldian social power reflects the rise of the prominence of the middle classes, the bourgeoisie, whose interests (and not those of the sovereign) begin to be represented by the practices governing this regime. The targeting of the "souls" of dogs, or the addition of dogs to the white man's civilizing endeavor, does not necessarily result from a recognition of their moral status. The vocabulary of gentleness is a façade; it is a cover up: a close analysis of the training apparatus reveals the regime to be deeply cynical at heart. However, it is interesting that the discrepancy between the training apparatus and the prevailing cultural definition of the human-canine relationship was not brought up – or was brought up only in relation to circus trainers – in the nineteenth century. After all, declarations of kindness clashed with the techniques of training proposed by the manuals. While kindness is, of course, a subjective category, the lack of any kind of opposition to the blatant incompatibility of the declarative level and the training apparatus is striking. The trainers speak of kindness and yet, inevitably, they all turn to the whip. However, let us not forget that within social power – just as within the disciplinary regime – the threat of penalty is a guarantee of the functioning of the system. Within these regimes, it would not be possible to conceptualize a training methodology that did not utilize punishment. Indeed, what begins to develop is "a sort of reasonable aesthetic of punishment" (Foucault 1978, 106), which embodies the predominant cultural values of the time: an appreciation of personal self-restraint coupled with a strong belief in civilizational progress.

CHAPTER 2

Hunting Dog Manuals: The Pointer as a Work of Art in the Age of Biopolitical Reproduction, 1845–1909

> Once the state functions in the biopower mode, racism alone can justify the murderous function of the state.
>
> MICHEL FOUCAULT, *SOCIETY MUST BE DEFENDED*

•••

> In "dem good ole days" when the plantation songs enlivened the labors of the hundreds of blacks as the golden harvests yielded to their cradles, and "ole marster" rode among them over his broad acres on his easy-going pacer he was almost always accompanied by a brace of thorough-bred pointers. The poor white man and the free negro hunted "ole hyars and 'possums' " with the hound; but the bloods of Virginia bird-hunted on thorough-bred horses and with dogs of pure blood likewise. With this picture of the past we still remember the bird dog as the Gentleman's Dog!
>
> C.A. BRYCE, *THE GENTLEMAN'S DOG*

∴

While this book is primarily about the training of pet dogs, by which I understand dogs kept in the house primarily for reasons of companionship, this chapter constitutes a necessary digression. It is a digression because it speaks of the training of hunting dogs in the nineteenth century, but it is necessary because it is this story that best articulates the point that the two poles of biopower: thanatopolitics and the more positive methods of biopolitical control can be aimed at members of the same species; that while some dogs are singled out to be trained; others can be singled out to be killed, and that the notion of trainability has historically served as a biopolitical caesura; demarcating two populations assigned to the opposite poles. It is also a necessary digression because it exemplifies further entanglements of breed and race in nineteenth-century US. It is not coincidental, after all, that the supposedly trainable dogs were owned

 | DOI:10.1163/9789004380295_004

by whites and the supposedly untrainable ones by black slaves. Being associated with the right kind of owners, and by extension with the cultural values and social status associated with that group, literally safeguarded a dog's life, even if it did not save him from the occasional physical pain of punishment during the process of training. The notion of trainability itself was constructed in a way that reflected dominant white middle-class values about what constitutes humanity: the ability to control one's impulses. Finally, it is a story whose significance rings true to this day, even if the protagonists have now changed with pit bulls replacing curs and coonhounds as the type of dogs labelled untrainable.[1]

Claire Jean Kim has written extensively on the cultural construction of the "pit bull" as black using the media accounts of the Michael Vick incident, a 2007 case in which the black quarterback was tried and sentenced for running an illegal dogfighting operation, as her primary sources (2015). In her analysis, Kim not only notes that the pervasive association of blacks with pit bulls works as a tool for animalizing the dogs' black owners – an element of the larger association of blacks with animality, analyzed in detail by numerous scholars – but also explores how in ritualized activities involving certain types of animals, the non-humans become symbols of their respective humans and the character flaws of the particular animal are transferred onto the owner (Kim 2015, 256–258). All of this is, of course, entangled in the longstanding association of black masculinity and animality (Kim 2015, 255; Weaver 2013, 693–694). Kim's analysis is reminiscent of Clifford Geertz's pioneering reading of the Balinese cockfight, where Geertz read the cocks as detachable genitals with a life of their own (Geertz 2005 [1972], 60). Nineteenth-century hunts with dogs were performances almost as elaborately ritualistic as the Balinese cockfight with the dominant styles of hunting associated both with different

1 Interestingly, pit bulls had for a long time been associated with white male owners in the American South. Harlan Weaver posits that the association of pit bulls with African-American males emerges only in the 1980s. Meanwhile, the racist slur most often associated with nineteenth-century minstrelsy, the coon, is a shortened version of the word racoon, an animal that was hunted only by African Americans, and the dogs used by them for pursuing racoons: coonhounds. It should also be mentioned that if we think of the two types of dogs: coons and pit bulls as representative of stereotypical readings of black men, we can trace a change in the construction of black masculinity through the change in these associations: from coon (hound) to pit bull. While coon hounds were certainly seen as impulsive and untrainable (incapable of restraint), they did not exude the kind of danger that the pit bull currently does. If anything, there was an element of comedy to the whites' perception of the cur's or coonhound's hunting style: chasing a racoon up a tree and barking. Ergo, and mirroring the recent readings in the history of the construction of black masculinities (Kimmel 2001, Bederman 2008), black masculinity was not fashioned as dangerous in the pre-Civil War era; rather as tragicomically flawed.

dogs and differently raced people. White landowners hunted "for sport" with their pointers and setters, while the black slaves who hunted with their curs were subsistence hunters: they needed the food. The association of slaves with "cur dogs" was then just as strong as the pit bull association is now and was similarly grounded: the dogs and the humans supposedly shared the same negative features: instinctiveness and lack of self-control. Both pit bulls now and curs then were deemed untrainable.

It is not coincidental that the cur is basically a mixed breed dog (Derr 2004, 150–175), and that the pit bull was never recognized by the American Kennel Club (Weaver 2013, 691). This story, after all, takes place alongside the rise of the discourse of breed. Dog breed, that is. The second half of the nineteenth century is a period when the general canine population starts to be affected by a shift in breeding practices: dog shows begin to operate as the dominant tool for the selection of so-called breeding stock. Control over sexual reproduction is certainly a staple of biopolitical practices and while human control of the process of breeding has always been an inevitable element of domestication, not just for *canis familiaris* but for all other domesticated species as well, what happened with dogs in the second half of the nineteenth century has generated overwhelming consequences even for the contemporary canine population: many of the dogs we interact with today have been shaped by the biopolitical forces that rose to power in the nineteenth century. The topic has been thoroughly researched; with Harriet Ritvo opening the discussion in 1987 through an analysis of the class-related aspects of the breeding of pet dogs in Victorian England. Ritvo argued that purebred dog ownership reflected the dog owners' upper-class aspirations. Other scholars continued probing the significance of this practice, and particularly its confluence in time with the rise of the importance of the notion of race in the human world and the emergence of scientific racism (McHugh 2004; Skabelund 2008; Rasmussen 2016). Claire Rasmussen adds that transforming dogs into "predictable, regulable creatures" could serve as a sign of benevolent human control of nature (Rasmussen 2016, 80–81) and – analogously to how eugenics functioned in the human context – it could also deflect anxiety related to "inappropriate" behavior by placing the blame on heredity and suggesting the possibility of eradicating threats to social order through the control of the breeding population (84). The turn to biopower in the government of human populations was accompanied, even predated, by the emergence of biopower as the predominant mode of managing populations of domesticated animals. In the context of canines, the notion of trainability played a key role in determining which dogs are made to live, which are allowed to die out and sometimes even, as sovereignty remains the underbelly of biopower, which are made to die.

While this book is primarily about training, the similarities in the goals of breeding and training in the second half of the nineteenth century sound uncanny: both practices aimed at forming a predictable, controllable dog, whose appearance and behavior reflected credit upon the animal's owner and/or breeder by confirming his social status and his/her involvement in the charitable endeavor of "civilizing the brutes." At the same time, the lack of understanding of the actual rules of heredity evident in the pet dog training manuals discussed thus far, resulted in the previously mentioned erroneous beliefs like telegony or the inheritance of deformity (a dog with an amputated leg seen as bound to have a litter of three-legged puppies). Not only were training and breeding involved in the same project, the effects of these two practices were conflated in the figure of "the genteel dog": the dog who is well-bred, beautiful and well-behaved; implicitly, because of his heritage. Genteel behavior, understood as self-restraint and control over one's animalistic impulses, was seen to result from proper breeding. In other words, dominant Victorian beliefs concerning human heredity were transposed onto animals (Ritvo 1987, 45–79) and breeders took on the challenge of introducing the Platonic ideal of the unity of truth, beauty and goodness into the canine population. As the last part of this chapter shows, particularly in the American context this ideal was highly racialized: it so happened that dogs considered both beautiful and trainable were those associated with and developed by white Europeans. The conflation of training and breeding, the belief that a well-bred dog will also behave well, was one of the major factors influencing the ideological confusion of training manuals for pet dogs: on the one hand, training was supposed to elevate the dog's mind (even that of "the meanest mongrel"); on the other, such elevation was the result of the work of breeders. This resulted in what I have referred to as the cynicism of the pet training textbooks: they promise the possibility of transcendence of the animal state and then do not deliver on the promise. They fundamentally cannot deliver because they are involved in a project that is both racist and speciesist to the core: it is an empty promise. While some dogs (purebreds) are better recipients of human civilizing efforts than others, no dogs can ever transcend their animality completely. Within the framework of the texts under discussion what they can do – if they are well-bred – is achieve ennoblement through gracious acceptance of their nature under the guidance of man.

It cannot be denied that in the case of the training of hunting dogs, heredity plays a major role. Ethologists now agree that it is thanks to selection for specific behavioral traits, such as the pointing behavior seen in many hunting dogs (most generally: pointers and setters) or the gathering and stalking behavior of herding dogs, that the hunting behavior of *canis lupus* was modified in a way that assured the passing on and strengthening of these desirable (for humans) behaviors across generations (Coppinger and Coppinger 2001). However, not all hunting dogs point. Some, mostly those we know as hounds and those jointly referred to as curs in the nineteenth century, identify prey, follow it (using either nose or eyesight), chase it and grab it. While the breeding of pointing dogs requires modification of the wolf's hunting behavior – the stopping of the locate > eye > stalk > chase > grab > bite > dissect > consume chain on the eye > stalk link – locating, chasing and grabbing fleeing prey are both instinctive behaviors that are part of the wolf's chain of hunting behaviors: eliciting them does not require much modification through breeding selection (Coppinger and Feinstein 2015, 58–59). As will become clear in the course of this chapter, in the social world of American sportsmen in the second half of the nineteenth century, those animals whose behavior was the most modified through human breeding selection (as a result: who were deemed the most trainable) were truly prized, while those who hunted in a no less effective but less modified way were looked down upon, identified with working class and African American populations and often seen as pests that had to be exterminated. In a nation bursting at the seams with racial tensions leading up to the outbreak of the Civil War and then not subsiding in the reconstruction era, the association of some hunting dogs with whiteness and others with blackness on the basis of their trainability, understood as susceptibility to civilizing human influence, is telling commentary on contemporary attitudes to race and class.

1 Sports Hunting

As mentioned previously, the publication of the first books on hunting dogs predates the boom on the pet dog training market by about a decade. While some books for hunters containing instructions on training were published in North America already in the eighteenth century, one of the first books specifically on the training of hunting dogs written by an American author for the American reader was published in 1845 in Philadelphia.[2] Already within its title J.S. Skinner's *The Dog and the Sportsman* (quickly followed by Elisha Lewis's book published in 1855 as *The American Sportsman*) contains the word that was making waves among American hunters at the time: the sportsman, who was also clearly the intended reader of publications of such type. This is not to say that using the word sportsman to designate the hunter is a nineteenth-century

2 Peter Hawker's *Instructions to Young Sportsmen*, originally published in the UK in 1830, was published by Blanchard and Lea in 1846 with the substantial addition of approximately 250 pages devoted to "American Hunting and Shooting."

phenomenon – there exist records that reveal that the word had been used this way in North America already in the eighteenth century – but it is in the nineteenth century that the designation becomes involved in a class/race struggle that separates those who hunt for subsistence and for sport. Writing on the basis of Alexander Hunter's memoirs, *The Huntsman in the South*, published in 1908 and covering the second half of the nineteenth century (from the antebellum period through the reconstruction era), Stuart A. Marks notes that sportsmen:

> tracked their quarries in a highly stylized manner, conversed with a technical vocabulary explicit about game and guns, demonstrated an abiding interest in natural history, adhered to a code of ethics, donned fashionable dress afield, often professed an interest in highly trained dogs, and belonged to cosmopolite associations. Only certain mammals and birds were worthy to test the skills of a sportsman and his dog. These prey were taken in a highly refined manner. Deer were given ground to exercise their natural instincts for escape, and birds were taken on the wing. Sportsmen, generally, were urban, wealthy, Eastern, and professional. (Marks 1991, 46)

Alexander Hunter himself wrote that when the sportsman "goes for a month's shooting he goes for pleasure, not for profit. (. ...) of course he rides; to tramp for hours on foot may suit a pot-hunter, market-gunner, or our African brother, but does not appeal to the amateur sportsman" (quoted in Marks, 45). Clearly, the dividing lines were wealth, class and race. Gender was an obvious given. While hunting in North America had been generally a much more democratic pursuit than in England due to the abundance of game, the second half of the nineteenth century is a time when game depletion, as a result of overhunting and increased industrialization, heightens the already existing tensions between the different types of hunters (Giltner 2008, 45–77). The books published in the second half of the nineteenth century were aimed at the sportsman, which explains their preoccupation with certain types of prey, certain types of weapons, the handling of horses and the privileging of certain hunting dogs over others. They also present hunting for sport as a nobler pursuit and one that expresses a concern with "nature"; in contrast to subsistence hunting, which was vilified as responsible for the gradual disappearance of game animals. The prey considered most suitable for sportsmen were birds: with quail and grouse leading the unwritten hierarchy. This is clear from the contents of most of the textbooks: Elisha Lewis's *The American Sportsman* (1855) already posits itself as a book that is specifically about hunting for fowl.

The reasons for the high status of game birds are manifold: even though mammals were still plentiful in most parts of the North American continent, hunting for birds was seen as more refined and sophisticated because of the activity's association with the English aristocracy. While foxhunting with hounds also seems to perfectly fulfill the same criteria, this analysis should prove that a different kind of a performance is generated by a hunt with setters and pointers than one with hounds: a spectacle that better exemplifies the key elements of the human-animal relationship of the period, specifically human influence on canine instinct.[3] Secondly, hunting for birds provided a more potent aesthetic spectacle, particularly the practice of "shooting on the wing," that is shooting a bird in mid-flight, advocated as a more sportsmanlike technique because it gave the bird better chances of escaping, but in reality providing an opportunity for an almost artistic display of skill in marksmanship. Thirdly, bird hunting could much more easily be turned into the competitive and sporty endeavor of field trials than hunting for big game, like deer, wolves, or even foxes. Finally, and most importantly, hunting for "vermin" (squirrels, raccoons and opossums) was firmly associated with subsistence hunters, who often did not even use guns. When white sportsmen engaged in raccoon hunting, which was considered a form of entertainment rather than hunting proper, the endeavor included the renting out of African-American dog handlers with their dogs. It would be the African Americans' dogs that would chase the raccoon up a tree and the dogs' handlers would then climb up the tree to capture the raccoon: thus, providing hilarious entertainment for the white sportsmen (Giltner, 59–67). John J. Audubon thus described the hunting of possums and in the American edition of Peter Hawker's *Instructions to Young Sportsmen*:

> As an animal of sport, the opossum is of course an inferior character; the negroes, however, look upon the animal as the most perfect game, and are much astonished the fox and deer should be preferred; and the hilarity with which they pursue the sport of 'possum hunting, far excels the enthusiasm of the most inveterate follower after nobler game. (quoted in Hawker 1853, 315)

2 The Notion of Breed and Hunting Dogs

Hunting was one of the earliest activities that humans and dogs engaged in together. However, the new notion of the "purebred dog," developed only in the nineteenth century, also impacted the world of hunting dogs in substantial

3 For an analysis of the contemporary foxhunt as performance see Garry Marvin's "A Passionate Pursuit: Foxhunting as Performance" (2003) and "Cultured Killers" (2001).

ways and at an important transitional period. While the idea that behavior is heritable was nothing novel for breeders of hunting and herding dogs, what was new – and somewhat confusing – was the need to figure out whether behavior is associated with specific physical conformation (not just with a general body shape but more specific points, such as eye color, head shape, white markings, etc.) and whether maintaining desirable behavioral traits requires the maintenance of a closed gene pool and a long pedigree. On the one hand, accepting these associations went against the lived experience of interacting with working dogs; after all, the eye color or size of a white patch on a dog's chest have absolutely no bearing on a dog's suitability in the field (albeit superstitions abounded among working dog breeders and handlers). On the other, it was tempting because the acceptance of such premises could raise the social prestige of the breeders and prove that they, too, were involved in the great project of "civilizing the brutes." In the developing tension between sportsmen and subsistence hunters, associating sport hunting with the world of purebred dogs could additionally imbue both with a range of desirable symbolic meanings: the prestige of sports hunting would increase through its association with the world of dog shows, while the prestige of purebred hunting dogs could also increase once they proved their worth in the field, not only in the show ring. When looking at the results of field trials and conformation dog shows in the last two decades of the nineteenth century, it is easy to see the overlap between the participants: it was desirable to own a dog that exceeded at both these activities.[4] Admittedly, rather than siding with lived experience (which would also be confirmed by science in the late twentieth century), the textbooks inevitably fell for the Victorian ideological project that training was embedded in.

The result is a certain level of confusion represented in the training manuals of the time. The textbooks invariably present themselves as allied with the emerging purebred dog fancy and most usually focus on a discussion of the various canine breeds. They often do acknowledge the desirability of a certain physical conformation for work in the field and the heritability of traits that most certainly, as contemporary science tells us, are not heritable. Many

4 It should be noted that the first Field Trial took place near Memphis, Tennessee in October 1874. This was still before the founding of the American Kennel Club (1884). As the century closed, the popularity of field trials increased. These events were organized in the Northeastern and Southern parts of the United States and – similarly to conformation dog shows – reports from the trials were published by major local newspapers and hunting magazines. The founding of *Forest and Stream* in 1873, edited in its first decade by S.T. Hammond, was particularly influential for the organization of the first field trial and created a platform for the publication of reports from field trials and for the exchange of opinions among "sportsmen."

American authors shared the English beliefs on heritability famously described by Harriet Ritvo in *The Animal Estate*. J.S. Skinner, the author of one the first hunting textbooks published in the United States (1845) was a deep believer in telegony (19) and advocated that a female dog should be hunted throughout her pregnancy because "pups of a well trained slut, and one that has been hunted during the greater part of gestation, are (…) better than others" (18). Many others explicitly shared the ideal of beauty and purity of breed influencing performance. For example, many authors believed that the shape of the dog's ears and head had a direct result on the dog's performance. Furthermore, hunting dogs were believed to be the most intelligent because of their elongated heads. As Frank Forester (pen name of Henry William Herbert) wrote in 1856, this shape "gives room for the capacity of brain, which, whether in man or the inferior animal, invariably indicates or produces superiority of intellect" (172). However, all textbooks acknowledged that in order to train a good hunting dog, one must begin with a dog bred from working parents. S.T. Hammond thus advises the reader in the process of puppy buying: "[i]n selecting a puppy there are many things to be taken into consideration. In the first place we must be sure that both sire and dam are first class field performers. This we consider of the utmost importance. They must also be possessed of endurance, and must be reasonably intelligent" (Hammond 1882, 2). In fact, all the authors are steadfast in this conviction. Waters adds: "[t]o attempt to make a dog into something other than he is naturally, is to engage in dismal efforts" (Waters 1889, 81).

Even though all the textbooks agree on the necessity of hunting heritage, there is confusion about whether the dog should be of purebred lineage or not. Most of the textbooks saw modern (purebred) hunting dogs as superior to those of the past because, through breeding practices, they have been influenced by culture and thus improved. Beauty seems to be an effect of culture, refining and polishing nature in the same way in which training refines and polishes the dog's natural instincts. For example, Forester writes that the ear of the setter "looks as if it had been rounded by art" and the dog is "perhaps improved by culture" (173). Also speaking of the setter, he adds that "[h]is beauty is a sign of the purity of his race; and in some sort (…) an indication of superior intellectual qualifications" (179). However, in the numerous anecdotes contained in the books, many of the authors recall that their best hunting dogs were not purebred. Some even acknowledge that in certain parts of the country the crossing of pointers and setters or pointers and spaniels is necessary because the pointer's coat is not suitable for cold weather and is not water-resistant. Towards the end of the century selected writers begin to show awareness of Darwin's *On the Origin of Species* (1859), where Darwin speaks of what is now known as inbreeding depression and hybrid vigor. After a thorough analysis of the pitfalls of inbreeding (the breeding of close relatives), Waters writes in 1889 that "while pure blood is indisputably a prime requisite in breeding, it can be rendered, by misuse, of less value than good mongrel blood" (371). In other words, the books reflect a tension between the general Victorian imperialist project, which precludes miscegenation (and which most certainly is also the predominant attitude to interbreeding among humans of different races in the US at the time), and the emergent Darwinian paradigm, which – even though it does not really pose a challenge to the idea of progress – identifies the pitfalls of closed breeding populations.

Still, even if some recognition of the practical value of mixed breed dogs was generally present within these textbooks even before Darwin's work became a clear influence on the thinking of the authors, there was awe and admiration for the purebred dog even if the purebred proved less useful. This was evident in the great retrieving debate that took place in the pages of *Forest and Stream* magazine in the year of the first field trial in the United States (1874). As the rules for the tests of a hunting dog's ability were being negotiated, a controversy emerged over the specific tasks to be performed by the pointers; specifically, the question of whether pointers should be asked to retrieve the bird after it had been shot. In England, where Field Trials began to be organized already in the early 1870s, pointers were not scored on their retrieving ability but only on the scenting and pointing of game, while retrieving was to be done by retrievers. However, some of the American "sportsmen" argued that they needed an "all around dog" who could retrieve as well as point. This desire seemed to echo the more democratic nature of hunting in America. As the Tennessee Sportsmen Association wrote to the editors of *Forest and Stream* in 1874: "To talk plainly, the majority of our sportsmen do not wish to be bothered, nor have the time, inclination or estate to keep a large kennel of dogs." However, the Association also understood that pointers, as highly bred dogs without a heavy protective coat, could easily take cold if they were asked to retrieve: "It is true that a high-blooded pointer will go anywhere, but not without ill effects to his person. (….) Their delicate organization, thin skin, slight coat, and having no hair between their toes prevents them retrieving successfully" (Tennessee Sportsmen Association 1874, 72). In the end, a compromise was reached with the longer haired setters being asked to retrieve game and pointers having only to "stand" their game: that is show the bird's location by pointing.

In other words, already from the very beginning of field trials, these competitive tests were not meant to evaluate the practical usefulness of a hunting dog, but to showcase a highly stylized manner of hunting, which was an aesthetic spectacle that could only be enacted and appreciated by those with the necessary cultural capital. The pointers and the setters were accessories but also embodiments of this aesthetics of hunting. Even if their practical use was

FIGURE 2. The pointer on point was a mesmerizing spectacle, reflecting the dominant values of the age: poise and self-control. New York: N. Currier.ca. 1846.
COURTESY OF LIBRARY OF CONGRESS, LC-DIG-PGA-09636.

sometimes of limited value, their symbolic value was paramount. Garry Marvin, analyzing contemporary English foxhunting (or almost contemporary: foxhunting was banned in the UK in 2007) speaks of the hunt as a performance governed by a certain aesthetics in which effectiveness is actually a secondary concern (Marvin 2003, 56). It is an aesthetics that is marked by "excess and intensification" (47), which in the case of the foxhunt consists of the elaborate rituals surrounding the practice, with the performance of the hounds – including their appearance, movement, the sounds they make while on the trail of a fox – constituting its key elements. In the case of the hunt with a bird dog the chain of events consists of: dog in motion/quartering > dog on point > hunter flushing bird[5]> hunter shooting on wing > bird falling to the ground (sometimes extended by: dog gently delivering shot bird to hand). The entire chain of events is imbued with a deeper symbolic meaning: it is a metaphor of the human-canine relationship in general. The aesthetic value of the performance is best concentrated in the image of the dog holding a point (Figure 2).

5 The issue of whether pointers/setters should also flush their game or only point/set it is another contentious one, very similar in its parameters to the debate on retrieving. Those who advocate against it, believe that teaching the pointer to flush "spoils" the dog and goes against his natural instincts.

3 Polishing Instinct: The Pointer as a Work of Art

The role of the training of hunting dogs would be to provide that polish of instinct seen as necessary if one is to talk of progress; and, after all, the entire Victorian enterprise was based on the idea of civilizational progress. This is why Forester expressed the belief that there was no training prior to the nineteenth century: in his eyes, training the hunting dog was a departure from relying simply on the dog's instincts. This was the quintessence of Victorian society: instinct needed to be controlled. This function of training is actually better visible in hunting training manuals than in pet training manuals for the simple reason that the former actually deal with the shaping of instinctive (or intrinsic) behavior. The behaviors performed by a dog during the hunt: quartering, pointing, setting, possibly also flushing, retrieving, have been shaped by the dog's genetic heritage. It is the putting of these behaviors under human control that is seen as the accomplishment of training. Prior to the nineteenth century, Forester sees the performance of a hunting dog as an expression of "pure instinct"; a term which could never serve as a compliment in the Victorian vocabulary. Accordingly, Forester goes on to contrast the backward views of his youth with the progressive, civilizational endeavor that training becomes by the mid-nineteenth century: "[i]t used to be held a maxim in my youth, that the dog of chase should be retained as much as possible a mere brute – that to cultivate his intelligence, nurture his affections, accustom him to understand your wishes and share your pleasures, was to unfit him for field service." Forester evaluates such beliefs as "brutal and barbarous errors" (155).

Being able to restrain one's instincts is seen as a mark of intelligence, and refinement. The reason why hunting dogs were considered to be superior in intelligence to others (with pointers and setters, as opposed to spaniels, retrievers, and scent hounds routinely listed as the most intelligent dogs) is because of their behavior upon the identification of game: they stop and literally point the game out to the hunter, but do not rush in on their own. It would seem that the expected behavior of a coon hound or a cur dog (using the vocabulary of these textbooks) would be to chase after potential prey immediately upon seeing it. The fact that pointers and setters become immobile and focus intensely on their object of desire without chasing and grabbing it, proved that they were capable of the kind of self-restraint usually associated with humans. Waters writes: "[t]hat both pointing and backing[6] may be exhibited as an act

6 Backing (also known as honoring) is the act of pointing upon seeing another dog point. Within the convoluted ideological framework of hunting, backing was seen as an act of honor. The dog could behave in an "instinctive" way and steal the prize (the bird) from

of intelligence, is established by many well attested cases, and is mentioned by several authors" (Waters 1889, 214). He then goes on to tell the story of a terrier (a dog of explicitly lesser standing among the writers of the time) who taught himself to point, thus proving his own intelligence:

> On the trail of a ruffed grouse, he was cautious, silent and attentive, which was entirely opposed to his manner of hunting squirrels, yet he learned both methods from his own experience. Here were precisely the same acts, exhibited by a bull terrier, which are commonly attributed to instinct in the pointer and setter, and which were performed as a matter of pure intelligence. (215)

The moral of the story is clear: even a dog of seemingly lesser intelligence – one prone to outbursts of instinctive behavior – may overcome that deficiency by learning to control his own impulsiveness. Of course, the story is one of the many anecdotes sprinkled throughout textbooks of the period; one in which an actual method for the teaching of the behavior is not provided: the assumption being that true intelligence cannot be taught.

The ultimate aesthetic spectacle in the hunt was the dog on point: body tense, bursting with desire and yet able to curb that desire, implicitly (though not necessarily accurately) because of his deference to his master. As *The New York Times* wrote of a field trial demonstration that took place at Madison Square Garden in 1878:

> The poor little quail[7] squatted down in the dirt, and about four feet removed all around stood in a circle of rigid, quivering dogs, every head thrust forward, every eye bulging with eagerness, every muscle set and every tail rigid. A prettier sight could not be imagined. A hundred

---

the dog on point. Backing was seen as chivalrous behavior and was actually a required element of a dog's performance in field trials. From an ethologist's perspective, there is nothing honorable in backing: it is exactly the same behavior, elicited by exactly the same stimulus, as the pointing behavior and one resulting from the same modification of the chain of hunting behaviors. The second dog does not stop because he honors the first dog; he stops because his hunting chain has been modified in exactly the same way as the first dog's.

7 It should be added that the phrase "poor little quail" is rather a cursory remark than an expression of serious pity for the birds. While sportsmen claimed that their hunting style of shooting "on wing" created a fair confrontation between hunter and prey, there is – quite unsurprisingly – hardly any concern in the texts from the period about the well-being of the pursued animals.

> spectators surrounded the ring gazing breathlessly at the picture, every countenance gleaming with delight. (….) One nervous young Irish setter could hardly contain himself, and was on the point of rushing in when a word from his master caused him to drop flat upon the ground where he wriggled from side to side in a feverish struggle between instinct and obedience. (*New York Times* 1878, 2)

The poise of the pointer on point was seen as mesmerizing, as the transcendence of animality (as the expression of ultimate self-control) and yet also its embodiment (all of this in the service of man). The hunting dog at point seemed to achieve a sense of dignity unattainable to any other dog: neither the pet dog doing tricks, nor the cur dog chasing a raccoon up a tree. This dignity was perceived as resulting from a graceful acceptance of the limitations that came with species. Only the acceptance of these limitations could lead to their final transcendence and instinct – or rather a winning struggle with instinct – was key in this achievement.

It is hard not to notice the implicitly racist undertones of such thinking: both cur dogs and African-Americans were seen as lacking in self-control. The transcendence of this shortcoming made the setter and the pointer truly aristocratic dogs, worthy of associating with the white sportsman. However, should a pointer break point, his fall was imminent. Stuart Marks quotes from Alexander Hunter's memoirs the story of an unruly pointer, one who broke point to grab the dead bird. Hunter advises the dog's owner to immediately discipline the dog: "I told him to cut a thick hickory switch and give the dog the worst hiding he ever had; that unless the vice of eating killed game was eradicated he might as well give him to some darky for a 'possum dog' " (Hunter quoted in Marks 44). It is not just the blatant racism of the statement above that is striking but the mechanics of the dog's association with the human. In a similar way in which the contemporary stigmatization of the pit bull is entangled in debates about race (Weaver 2013, Kim 2015), the possum dog, the coon hound, or the cur, served as a transposition of the supposed vices of African-Americans onto the animal world. When Claire Kim analyzes the pit bull's fight in the pit, she is interested in the question of why the losing dog must be killed after the fight is over and reaches the conclusion that among the "dogmen" – as pit bull handlers like to call themselves – the greatest vice is weakness (Kim 2015, 258). In a context in which the dog serves as a stand in for his trainer, it is logical that a dog that has shown weakness by losing a fight needs to be "disposed of" if the owner is to maintain standing within his peer group. Somewhat ironically, contemporary dogmen have adopted the derogatory term cur to refer to dogs that do not exhibit proper gameness; the ability to fight until death: "Killing curs quickly and brutally helps to alleviate the owner's humiliation and restore his masculinity" (259). In a similar vein, Erica Fudge – though speaking of pets, not of hunting dogs – suggests that a poorly trained dog reveals the limits of the owner's hold on power (Fudge 2008, 22–23). If a man cannot control a particular behavior in his animal, then his ability to do so with regards to himself becomes questionable. In the mid-nineteenth century culture of white sports hunters, the greatest vice was impulsivity and the dog guilty of it was deemed worthy only of being a "darky's" cur, or possum dog if the owner was to maintain his position among his peer group. Meanwhile, self-control was exemplified in the behavior of pointing.

There was some dispute among the writers about the origins of pointing behavior. Some, understandably, could not believe that behavior proving such higher intelligence could have developed among canines without human intervention. Forester was quite certain that pointing had been taught to a dog by a human and later passed on across the generations: "the habit of setting or pointing its game, which is now an instinctive and natural qualification of its race, was originally an acquired trick, taught by diligent breaking. Centuries of tuition have rendered that acquired trick an hereditary gift" (Forester 170). Forester's beliefs make it easier to understand the confusion present in the pet dog training books of the period as related to the heredity of acquired behaviors. As discussed in the previous chapter, it was quite a common belief that the ability to learn tricks, such as standing on hind legs, was passed on "from sire to son," even though this belief went against the practices of, for example, circuses that collected stray dogs from the villages they visited. The belief that behaviors acquired through the process of training could be passed onto the dog's offspring was yet another element of the set of beliefs about human superiority over the animal world. In all fairness, not all trainers shared that belief. Waters, the only one of the hunting book authors who wrote his treatise after reading Darwin, is much more skeptical about this possibility:

> The conjecture that at some uncertain period in past ages, in some vague manner, the dog's hunting and pointing were matters of education, should not be set against the fact of any individual, natural imperfections in the present. There is neither sense nor necessity in straining after far-fetched theories, to prove that the pointing instinct had an educational origin. (Waters 1889, 73)

It should be added that contemporary ethologists are quite unanimous about the origins of the pointer's pointing behavior: it is a modification of the wolf's hunting behavior in which the behavior chain of eye > stalk > chase > grab >

bite > dissect is modified in a way that makes it end with stalk. The motor pattern of chase never appears. This is accomplished through breeding selection of individuals that exhibit the desirable patterns (Coppinger and Coppinger 2001; Coppinger and Feinstein 2015). In a way then, there was a grain of truth, in Forester's beliefs: this specific motor pattern had been strengthened through human intervention in canine reproduction, even if it was not directly an acquired trick passed on "from sire to son." However, in the case of training dogs for tricks such as walking on hind legs, from an ethologist's point of view the belief in the heredity of such behaviors would be completely erroneous as in this case we are not dealing with modifying motor patterns that result from the dog's evolutionary heritage.

The recognition of self-control as a value in the American society of the late nineteenth century is also reflected in the training methods themselves. This is visible even more clearly in hunting dog manuals than it was in pet dog training manuals because hunting was indeed about instinct and passion, both for humans and for dogs. If we think of the training of the hunting dog, as the books of that time did, as the inculcation of a certain measure of control over instinctive behaviors, the purpose that the process of training serves for the trainer should be similar: it should teach the trainer to control his emotions while administering punishment. Many of the books have the greatest contempt for those who punish the dog in a fit of rage; they are seen as stooping down to animality as a result of the process of training, rather than as raising the animal's mind above the purely instinctive. In other words, progress is measured not so much in the decreasing severity of the punishment but in its thoughtful and controlled administration. Thus, training becomes an exercise of self-control for the trainer. Interacting with the dog's animal nature, with the dog's hunting instincts, pushes the trainer to the edge, forces him to come into contact with his own animal urges. Yet, he must learn to control his temperament to prove his own humanity. As Waters wrote in 1889: "[b]y self-discipline a trainer can learn to control his temper; if he has not the willpower to so control it, he is not the proper material for a trainer. Just so often as he loses his temper and inflicts unnecessary punishment, just so often does he do wrong and obstruct progress" (63).

At the same time, the techniques for inflicting punishment are certainly severe from today's perspective. Waters goes on to advocate the spike (prong) collar together with the whip on larger dogs because of the possibility that the spike collar alone may injure the handler's arm when dealing with a strong canine (86–87). In accordance with the idea that training a hunting dog is about curbing his instincts, most of the textbooks focus on deterring a dog from undesirable behaviors, such as: not stopping, rushing in to retrieve a bird upon hearing a gun, chasing game and running away from the trainer, refusing to release retrieved game, mouthing the dead bird, etc. Most of the techniques are based on more or less ingenious ways of inflicting punishment upon the dog at the time the "offence" is committed. In the case of hunting dogs, this can be quite a challenge as dogs generally work far away from the handler. Thus, an entire apparatus of training lines, check cords and other devices such as logs tied to a dog's collar are discussed. It will later become clear that the predominance of punishment-based methods was not an exception but rather a persistent pattern in the training of working dogs: hunting dogs, herding dogs, and later police dogs, as discussed in chapter 3, and that the persistence of punishment has to do with the implication of these activities in a framework of eugenic biopolitics. If the purpose of training is to select the dogs who exhibit a superior behavioral conformation, the dogs' capacity for withstanding punishment and reacting to it need to be tested. Only dogs who react properly are selected for further breeding. In this framework, evaluating trainability necessarily includes establishing the dog's reaction to punishment, and training that does not include "corrective measures" would not serve the purpose of ensuring proper breeding selection.

4 S.T. Hammond's *Training or Breaking?*

In the late nineteenth century, the one significant exception when it comes to the prevalence of punishment-based methods among hunting dog trainers of the period was S.T. Hammond, the author of *Practical Dog Training: Or Training vs Breaking* (1882). Not only did Hammond almost completely forego the use of what would later be defined by Skinner as positive punishment – that is the addition of an undesirable stimulus to stop an unwanted behavior – in training, he focused on using positive reinforcement for fostering desirable behaviors. In all fairness, it must be added that he did use the whip if the dog "willfully disobeyed" but only after "the pupil's education is complete" (45). However, what brings Hammond closest to many contemporary behaviorist trainers is the fact that his goal in using positive reinforcement was not just to elicit or lure a desirable behavior – such as by dangling a piece of meat above a dog's head to encourage him to stand up on his hind legs, as had been the case in many of the pet training manuals discussed in the previous chapter – but to develop the dog's self-control. In other words, while he most certainly saw the process of training as the shaping of a dog's instincts, the process was not based on physical curbing of these instincts by the human but on instilling self-control in the dog through methods that did not rely on punishment. This approach will later (since the late 1980s and into the twenty-first century) become the trademark of a new school of trainers, most often referred to as positive trainers, largely influenced by Skinner's behaviorism. Hammond is not just more "progressive" in his gentle methods; his training philosophy seems contrary to the philosophy of the day because he clearly does not share the belief in the redemptive power of punishment that I have presented as characteristic of the modality of social power. His approach is not one where the subject of training is saved from his own animal nature through the timely application of punishment – as in Alexander Hunter's advice to the owner of the pointer who broke his point – but one in which the subject develops an understanding of how his own instinct works and how it can be overcome.

This difference of approach is best visible in Hammond's procedure for training the stop behavior or "To Ho," where he proposes the technique that is known among today's trainers as "doggie zen," or "impulse control," allowing the dog access to the desired object (morsel of food, in this case) only once the dog relinquishes the struggle to reach the food himself. The dog learns that stopping when he sees something he wants is the only way of obtaining the treat. At no point in the training process does Hammond jerk the dog's leash or use a whip (which Waters does, rather judiciously, in teaching the same behavior). The one distinction between the contemporary exercise of "impulse control" and Hammond's "To Ho" is that in the case of the former, the dog is expected to focus on the trainer's face. Hammond teaches the dog to focus on the piece of meat but only approach it when permission is given:

> You should begin by giving him a taste of a piece of meat, then secure a firm hold upon his collar, and place a small piece upon the ground in front of him. He will struggle with all his strength to get at it, but hold him steadily, and do not say a word until he becomes partially quiet. (....)
>
> After a few seconds the dog will become more quiet, and you can repeat the words. Now carefully watch him, and as soon as his attention is fixed upon the meat, and he looks at it steadily for a second, release your hold and cluck to him as a signal that he can now have it and at once praise and pet him, and give him to understand that he has done something wonderful, and that you are pleased with him. (5)

All of Hammond's methods are surprisingly "modern," but they are compatible with his overall training philosophy. He teaches the drop of the retrieved object beginning with the exchange of a less desirable object for a more desirable one, a piece of "tough beef" for "liver" (8); he breaks bird dogs of pointing rabbits by raising them with rabbits, exhibiting a surprising understanding of what modern ethology calls accommodation in behavioral development (Coppinger and Feinstein 2015); he associates the sound of a gun with food, showing a good grasp of Pavlovian classical conditioning years before Pavlov's work became widely known. He teaches puppies when most other books of the period acknowledge that training should begin only once the dog is at least one year old. He teaches at home (or "in the yard") and then transfers these behaviors onto fieldwork, a practice that is commonly referred to as ground work or flat work by contemporary trainers of sports dogs. All of this goes against the grain of popular instruction of the period. For example, Bogardus writes in 1874 that "The field, where there are birds, is the place to break dogs, and puppies are too playful and too soft for the real breaking. At about a year old the dog is of an age to understand what is wanted of him in a short time, and also fit to endure the correction which will be required to make him avoid faults" (289).

Hammond's approach throughout the entire book is exceptional for the period: rather than positing the process of training as an act of human benevolence while proposing increasingly more complex methods of punishment for misbehavior, he refrains both from using the civilizational rhetoric and the whip. His advice does not fit into the broader framework of dog training at the time, which explains why his methods quickly went into oblivion, only to be recovered (though with no credit given to Hammond) in the late twentieth century. What is also unique is his analysis of the human-canine relationship in terms of love and reinforcement rather than through a hierarchy of creation and the dog's innate desire to please. Love is a very rare word in training manuals of the time and yet Hammond uses it: "Be very gentle with him at all times; carefully study his disposition, and learn all of his ways that you may the more readily understand just how to manage him. You should be in perfect sympathy with him and humor all his whims and notions and endeavor to teach him that you truly love him" (4). However, even for Hammond considerations of performance are of paramount importance, and the strong bond with the human, established through a history of positive associations, is expected to aid the dog's performance in the field. Should the dog's performance not prove excellent, he is to be disposed of. To Hammond's credit, he does mostly speak of selling puppies that do not show enough promise: "There is one thing that we consider of paramount importance – our pup must staunchly point when he is from six to ten weeks old. If he will not do this naturally and of his own free will, quickly dispose of him to some one who is not so particular, and try again" (16–17).

What is striking about all textbooks directed to the "sportsman" – and Hammond's book is one of them – is that they address the trainer/sportsman as a breeder. The way of correcting major faults in a dog is to preclude that dog's reproduction, to cull it, that is remove it from the breeding program. Minor faults are corrected via the selection of complementary mates. Training is the process through which one learns about the dog's strong and weak points, but it cannot be the primary corrective measure. Training here is a significant element of the biopolitical management of the canine population. It is not an end in itself; neither is it, or could it be, about nurturing and developing the individual relationship between a human and a canine. The relationship, though it most certainly does develop, is subjugated to the broader selective framework. Whatever surplus dogs are produced, need to be disposed of. From this perspective, the biopolitical management of hunting dogs is based on exactly the same premise as that of show dogs: identify faults and attempt to overcome them in the next generation. The faults are, of course, different but both these approaches depend on the expendability of canine bodies; those who do not meet the high standards are culled. While the word cull (from the Latin *colligere* – to collect) does not mean to kill but to remove from breeding, historically the process of culling has indeed included killing; even George Washington's strategy of dealing with hound puppies that did not meet his standards was to drown them (Derr 2004, 77). While in the late nineteenth century many breeders/trainers may have felt that their criteria for selecting breeding animals were much more superior to those of dog show devotees – and they most certainly did, as evidenced in the scathing irony of *Forest and Stream* authors when they wrote about ladies' lap dogs – the reality is that the structural parallels between the two practices are striking. One may claim, as the trainers of the period did, that selection based on performance is superior because it is based on the dog's usefulness to humans. Waters even claimed that breeding selection based on performance in field trials imitates Darwin's survival of the fittest. However, not only was such human-directed selection a highly intensified version of natural selection, it was filtered through the human's, often erroneous, assumptions about heritability and correlation between behavior and phenotype.

Finally, the emergence of competitive canine events – in addition to field trials, sheepdog trials began to be organized in the UK in 1873 and in the US 1880, while competitive sled dog racing took off in the second decade of the twentieth century – brought with it a change in traditional methods of breeding selection: a definite intensification of human control over canine reproduction and selection for traits that may not have been crucial in the period before the competitions began. The events that constituted the tools of breeding selection did not truly test "natural" ability but, as the above analysis of field trials shows, what they actually tested was the dog's compatibility with the aesthetic spectacle of the hunt. Field trial competitors selected for dogs who would be the most stylish hunters, not necessarily the most effective. Sheepdog trials, while outwardly aiming to *preserve* the *natural* talents of dogs, not only appeared at a time when the skills of sheepdogs were in lesser demand but also led to a shaping of the dogs' skills. "Style" became a primary concern for the triallers in a way in which it never was for farmers and yet it was "style" which naturalized the dog's performance, by making it look "smooth," "graceful," and "natural." Similarly to field trials in both the US and the UK, early sheepdog trials in the UK were, as historian Albion Urdank traces, "gentry sponsored and therefore heavily freighted with elite values concerning the shepherd's craft. Early trials accented speed, agility, and obedience in the dog, as well as sheer entertainment value" (Urdank 2006, 65). In fact, concluding an ethological analysis of the behavior of contemporary border collies, Lord, Schneider and Coppinger add:

> The one caveat to the above discussion that should be mentioned is that the herding behavior illustrated is that of trial dogs. As with the sled dogs, the beginnings of competition trials changed the nature of the dog. It is doubtful that the trial performing dog would be of much use to the majority of pastoralists nor would the average shepherd have the time, skill, or inclination to train such a dog. (Lord, Schneider and Coppinger 2016, 52)

5 Hunting in Black and White

However, to return to the mid-nineteenth century in the United States, the very existence of the manuals for the training of hunting dogs, which are clearly involved in advocating a "proper" way of hunting, covers up the existence of a counter-tradition against which the sportsmen position themselves: that of subsistence hunters. The methodology of this book, my focus on an analysis of published manuals, takes the gist of my research away from those who did not have the means to write and publish books. Yet, it needs to be emphasized that during the time period under discussion dogs were most certainly trained to hunt in a style very different from that advocated by proponents of field trials. The story that seems most potent as something of a counter-history – and one which I can only briefly recount on the basis of already published research by historians, mostly works by: Stuart A. Marks (1991), John Campbell (1994), Scott E. Giltner (2008) and journalist Mark Derr (2004) – is that of slaves and their hunting dogs in the last decade of slavery and in the South under reconstruction.

In this story dogs became an important medium through which the tensions between master and slave were enacted (Campbell 54). All of the authors mentioned above insist that slaves depended on activities such as hunting to withstand bondage, specifically to supplement their diet and provide meat for their families. It is also clear, both from the accounts of former slaves (Giltner 1–2) and from the accounts of white landowners, that they hunted using dogs. In fact, they relied on the dogs to a far greater account than wealthier whites who had easier access to guns. Slaves' dog had to chase prey into a dead-end location, or up a tree, and were then either called off by their owners or actually made the kill on their own. Of course, slaves also used other hunting techniques that did not involve dogs, such as setting traps and hunting with fire, but this does not detract from the importance of the dogs. For the slaves, dogs not only provided hunting partners but also cherished companions (Giltner 1–2, 23–25; Campbell 54) and reminders of their own humanity. As Campbell writes: "The very fact of having a dog represented a vital assertion and demonstration of slave humanity in defiance of what slaveholder law and daily practice proclaimed to the contrary" (56).

While records of actual training techniques are very scant, it is clear that slaves were proud of their dog training skills. Giltner cites a conversation recorded by Harry Worcester Smith, a Northern sportsman who travelled through the postbellum South on hunting expeditions, with an African American servant who recalled how his father would train his own dogs: "He told me the names of all the spaniels and how three of them would 'fetch' but he 'spected' the two younger ones would soon learn, that his papa could make a dog do anything" (Smith quoted in Giltner 24). Antebellum sporting narratives also abound in mentions of slave dog-handlers on plantations. These handlers usually managed and trained packs of foxhounds belonging to their white owners (Giltner 91). The ubiquity of African American hound handlers in the antebellum South brings to mind the previously discussed distinction between the different types of hunting dogs and the different types of hunting. Hunting with foxhounds was certainly an activity which bore a resemblance to the lifestyle of the English elite, who also kept special handlers to care for and train the dogs. At the same time while the activity itself was elitist, the training of the hounds could be relegated to slaves because it was not seen as requiring much sophistication or intelligence. On the other hand, hunting with bird dogs was seen as requiring true skill and a more hands-on approach by the dog's owner, not something that could be left to the care of a lower-class (or Negro!) servant. However, the dogs most African Americans were associated with were not spaniels and foxhounds but so-called cur dogs, or coon hounds: dogs whose breeding was certainly not "pure" and yet who managed to hunt successfully for their owners. Marks writes that slaves "participated as drivers and subordinates for their masters during the daytime, and at night went after raccoons, opossums, and other game with dogs" (28).

Ownership of dogs by slaves as well as their nighttime hunting exploits were generally seen as a nuisance by slaveowners. The dogs themselves were routinely described as undernourished, ugly and positing a health hazard for the plantation. Most significantly, slaves' dogs were blamed for killing sheep belonging to the white plantation owners (Campbell 64–65, Derr 84–85). Such complaints had been submitted ever since sheep-raising became a significant factor in the economy of the American South, and they were usually resolved by individual plantation owners who periodically ordered the killing off of slaves' dogs. Such a solution had even been implemented by George Washington himself (Campbell 67, Derr 84–85). However, it is particularly poignant that these complaints escalated shortly before the outbreak of the Civil War, with pressure mounting toward the banning of dog ownership by slaves. Virginia passed legislation mandating physical punishment for slaves caught hunting with their own dogs. Significantly, these same men were often handlers of their owners' dogs: no persecution followed should they be discovered carrying out an errand for their owner. Similarly, slaves were banned from travelling with dogs, under the assumption that such dogs could kill sheep. Again, if they were travelling with their owners' dogs, the provisions did not apply "the argument being that those well-bred animals did not kill sheep" (Derr 85). In 1859, rather than banning the possession of dogs by all African Americans, the state of South Carolina made dog owners liable for sheep killed by them. Additionally, the legislature adopted a tax of one dollar for every dog owned by a slave (Campbell 67–68). This fiscal burden was too heavy for the slaves. As Campbell writes, the passing of these measures led to the decimation of the canine population and would have led to the total extermination of all dogs owned by African Americans in South Carolina if only the Civil War had not interfered with the implementation of the law.

The fate of slaves' hunting dogs in the late 1850s was supposed to serve as a reminder of the status of the slaves themselves: they were chattel to be disposed of freely by their masters. It was supposed to serve as a warning for the slaves; hitting them hard both economically (without dogs, they could not supplement their diet) and emotionally. If dog ownership served as a mark of humanity for men (literally not metaphorically; it was mostly men who owned hunting dogs) who were devoid of the legal possibility of establishing a family, taking their dogs away was a particularly cruel reminder of their status. Yet, this was not the taking away of the dogs. The euphemism stands for their killing. This was an expression of whites' power mediated via the killing of actual canine bodies. The dogs served as stand-ins for the slaves because they had long become associated with the "bad" features of the Negroes: their supposed impulsiveness, inability to overcome their animal nature, innate thievishness. All the sophisticated biopolitical strategies for improving the qualities of certain canine populations were coupled with the killing of those deemed irredeemably corrupt.

To bring the notion of training into this debate, such acts could have been excused through the framework of trainability. The assumed problem with the slave dogs was that they, unlike the white hunters' pointers and setters, could not be trained to control their wolfish sheep-killing nature. Unsurprisingly, bearing in mind Foucault's notion of the docile body, trainability was thus a criterion for simple survival. Of course, such an explanation served as a smokescreen. These dogs were killed not because they were not trainable, but because they were owned by the wrong people. The white slave owners could not target the slaves directly because slave labor was key to whites' economic prosperity. In something of a mirror reflection of Darnton's *The Great Cat Massacre and Other Episodes in French Cultural History* (1984) – when French apprentices, unsatisfied with their work conditions and the unjust treatment they were receiving from their masters, killed their masters' cats to distress the masters – animal bodies once again served as scapegoats in a conflict between two groups of humans. However, the excuse of "untrainability" only makes sense in a biopolitical mode, where training and breeding are commonly accepted as part and parcel, as inseparable. Furthermore, the dog's refinement through breeding, the number of generations of purebred ancestors in his pedigree, is seen as indicative of the animal's trainability.

While these episodes could be read as a revival of sovereign power, this was not the case. They were actually highly typical for the emerging modality of biopower. While it is Agamben's reading of biopolitics that emphasizes its easy slippage into thanatopolitics, Foucault also notes that the emergence of biopower was coupled with extreme acts of violence toward certain populations, though he most certainly is not referring to animal populations. It is racism that makes such violence possible: the shaping of certain populations through "positive" biopolitical measures is coupled with the eradication of others, all in the name of racial purity. In *Society Must Be Defended*, Foucault observes: "If you are functioning in the biopower mode, how can you justify the need to kill people, to kill populations, and to kill civilizations? By using the themes of evolutionism, by appealing to a racism" (Foucault 2003, 257). In *History of Sexuality. Volume I* Foucault observed that contemporary genocide (most obviously: the Holocaust) does not signify a return of sovereign power but happens "because power is situated and exercised at the level of life, the species, the race, and the large-scale phenomena of population" (Foucault 1978, 137). In *Society Must be Defended* Foucault further explores this dark underside of biopower, its seamless transition into thanatopolitics, especially in the Nazi regime, a theme later taken up by Esposito and Agamben. Thinking of Nazi genocide, Foucault writes: "The two mechanisms – the classic, archaic mechanism that gave the State the right of life and death over its citizens and the new mechanism organized around discipline and regulation, or in other words, the new mechanism of biopower – coincide exactly" (Foucault 2003, 259). He continues this argument, adding that "Nazism alone took the play between the sovereign right to kill and the mechanisms of biopower to this paroxysmal point. But this play is in fact inscribed in the workings of all States" (260).

Such observations of the workings of biopower have already been applied to the animal world. For example, James Stanescu writes of animals in contemporary factory farms as the "supplementary inverse of biopolitical logic" (Stanescu 2013, 135). Stanescu further elaborates on this theme:

> There seem to be two co-supplementary movements of the biopolitical, both inherently violent and imperialist: one is a move of rendering someone as inhuman and therefore disposable, and the other is a move of rendering someone as human and, therefore, in need of training and eradicating all traces of the inhuman. These moves are not oppositional discourses; instead the two notions of the biopolitical merge (or are better understood as connected through a linchpin) into that of the thanatopolitical. (141)

This duplicity of biopolitical logic is clearly visible in the historical episode from the 1850s, although the episode under discussion is even more complex because of the use of one species (even one population within it) as a stand in for a population of humans. Stanescu's use of the word training coupled with its spatial and definitional proximity to "eradicating all traces of the inhuman" perfectly renders the logic of the processes at work in the American South in the 1850s in relation to hunting dogs. To be (worthy of being owned by a white) human is to be in need of training. It becomes clear that trainability – understood as the animal's capacity to embody notions of phantasmatic dogness, a dogness that at once transcends animality but also embodies it – serves a supplementary explanatory function in this racist framework: those rendered trainable, capable of becoming docile subjects of the biopolitical state, are subjected to "positive" biopolitical technologies. Those rendered untrainable may find themselves subject to technologies reminiscent of sovereign power but applied on a massive scale: technologies that are much more efficient than the technologies of sovereign power. It is not coincidental that in the mid-nineteenth century the valorization of the pointer and the breed's selection as a subject of training was coupled with the mass extermination of slaves' coonhounds. These two processes are both elements of the framework of biopower. Speaking of the rise of scientific racism in the nineteenth century, Claire Kim writes: "If the hallmark of humanity was perfectability, progress, and history, Negroes seemed to lack it" (Kim 2015, 37). The Negro, much like the coonhound, was deemed "untrainable." To use Giorgio Agamben's terminology, the distinction between *bios* and *zoë* is here drawn within one species through the category of trainability. The arbitrariness of this category and its implication in a racist framework continues to haunt Americans' interactions with dogs to this day.

The analogies in the stories of nineteenth-century coonhounds and curs and contemporary pit bulls are striking. Claire Kim argues that "Pit bulls are dying for being black" (Kim 2015, 273) and reads the history of BSL (breed-specific legislation) as a modern-day racism, in which the dogs are discriminated against or simply sentenced to death – there exist many shelters that do not even try to adopt out pit bulls but routinely euthanize them upon admission – because they are *a priori* judged as untrainable, incapable of rehabilitation, much like the people they are associated with. The conclusion of the story of Michael Vick and his pit bulls can, however, be read as offering a hope of transcendence of the logic of biopolitical framing: the dogs seized on Vick's property were assessed by canine behaviorists, and many were rehabilitated and rehomed, becoming cherished and loved family pets. After serving time in prison, Vick himself has become involved in animal protection. The story of the former fighting dogs turned pets has gathered almost as much attention as the controversy surrounding Vick's illegal fighting operation. Several tear-jerking books have been created about the dogs' lives after their release from Bad Newz Kennels: Jim Gorant's *The Lost Dogs: Michael Vick's Dogs and Their Tale of Rescue and Redemption* (2010) and *The Found Dogs: The Fates and Fortunes of Michael Vick's Pitbulls, 10 Years After Their Heroic Rescue* (2017); a documentary following the lives of the individual dogs, titled *The Champions*, was also produced in 2016. All of these texts present the dogs as transcending the label of incorrigibility, changing, adapting and progressing under the influence of a combination of training and love. They are certainly not untrainable; in fact, the opposite proves to be the case: it is their malleability that we begin to admire; their ability to go from being a fighter to being a lover. As the subtitle of Gorant's first book suggests, the dogs' stories begin to circulate as narratives of rescue and redemption; of the power love. The question is whether the ready incorporation of the dogs into such a different mythology is an example of the transcendence of the duplicitous logic of biopower; of its transformation into a completely different mechanism in which all (members of a species) are equally worthy; or simply a replaying of the dynamic described in chapter 1: the "salvation through love" that seemed so dangerously close to "civilizing the brutes." To phrase the question another way, is the rehabilitation of the pit bulls an example of humanist (and humanitarian) logic, in which the frame of what counts as *bios* is extended onto yet another group (pit bulls) but the caesura (here: of trainability) remains? Or is it an example of a posthumanist approach in which this logic is done away with altogether? I will try to explore these questions in the chapter related to contemporary training approaches, but first I need to turn to a different moment in the history of human-canine relations; one which no longer valorizes the dog's malleability, sagacity and natural submission, but begins to emphasize the dog's latent wildness.

CHAPTER 3

Culture of Instinct: Emergence of the Disciplinary Regime, 1910–1946

> Kant defines two essential conditions under which mankind can escape from its immaturity. And these two conditions are at once spiritual and institutional, ethical and political. The first of these conditions is that the realm of obedience and the realm of the use of reason be clearly distinguished.
>
> MICHEL FOUCAULT, "WHAT IS ENLIGHTENMENT?"

•••

> The question whether or not the dog employs intelligence and rudimentary thought in any human sense is not relevant to practical training. Experience in thousands of cases has proved that it is only training on a memory basis that can lead to reliable service.
>
> KONRAD MOST, *TRAINING DOGS: A MANUAL*

∵

The dates that delimit the scope of this chapter represent the publication of two volumes that mark the symbolic beginning and end of an era: Konrad Most's *Dog Training: A Manual* (in German: 1910) and Blanche Saunders's *Training You to Train Your Dog* (1946). Most was the forerunner of the use of dogs by the police, while Saunders taught the first obedience class open to mixed breed dogs. Briefly put, Most advocated teaching dogs how to attack criminals, while Saunders taught pet dogs not to bite. This rather significant difference is representative of two projects that seem fundamentally incompatible and yet which are closely connected through the framework of discipline. The fact that the discussion closes with Saunders, whose writings are then analyzed in chapter 4, clearly points to which project won: the project of engaging American pet dog training in the techniques of governmentality; of using training as a means of achieving non-violent human-canine coexistence; of fostering a feeling that dogs are fundamentally companions rather than potential aggressors. By 1946

 | DOI:10.1163/9789004380295_005

training dogs to attack had become firmly associated with the uniformed professions (in which case it was acceptable) or lower-class and immigrant men (in which case it was not). Still, even displays of aggression by dogs handled by officers in uniform were unwelcome and a cause for public concern. When in 1963 the police used dogs against the peaceful protesters in Birmingham, public outrage ensued. In fact, the famous photographs showing German shepherd dogs attacking African Americans are sometimes said to have swayed public opinion in the direction of the Civil Rights movement (McWhorther 2001, 323–337). The fact that they did – or even that they potentially could have – shows that in 1963 attitudes towards public displays of canine aggression were negative. Here, however, I focus on America's brief love affair with the concept of the personal protection dog; the ensuing correspondence between the function of training and the methods used, and with the change in ideas on canine instinct and intelligence that facilitated such training.

This period is not fondly remembered by most contemporary dog trainers and is usually referred to as the "dark ages" of dog training for its propensity towards confrontational and force-based methods of training and its reliance on the choke and prong collars as required training equipment. However, this is a simplistic view that obscures two features that I see as key for interpreting training activities as both constitutive of and reflecting bigger cultural discourses. Broadly speaking, it is not just violence that is the standout feature of this period. The common denominators for many trainers were: revaluation of instinct coupled with the devaluation of canine intelligence and a deliberate move away from the kind of anthropomorphism that was typical of the second half of the nineteenth century. Significantly, at the time when the use of dogs by the armed forces and the police was becoming commonplace, the inclusion of canines in the nation-building institutions influenced the methods of pet dog training in a way that emphasized physical confrontation. At the same time, the police and military institutions provided a model on which mass-scale pet dog training could later be based. The almost metaphysical dimension of punishment typical of this period is actually reminiscent of the training philosophies discussed in chapters 1 and 2, although this metaphysics becomes even more developed as the practical apparatus for inflicting punishment evolves toward greater sophistication and efficiency: the whip – while not completely eliminated – is gradually faded out and training collars are championed as a more humane alternative.

In fact – in a by-now familiar strategy – all of the trainers whose books are discussed in this chapter routinely described their methods as humane. And while one could argue that they were cynical in their self-depictions, I do not think this is the case: they genuinely believed in the necessity of physical punishment – the disciplinary regime is inconceivable without at least the threat of punishment – and in the need for establishing a human-canine hierarchy through physical confrontation. As will become clear, pet dog trainers incorporated the notion of usefulness into the rhetoric of training pet dogs, in an attempt to raise the prestige of the activity and signal its seriousness by connecting it with the budding use of dogs by the uniformed professions. Konrad Most is quite clear about the object of training: "It is that the dog shall only do what we find convenient and useful, and refrain from doing what is inconvenient or harmful to us" (Most 1955, 12). Of course, what humans find useful depends on their individual circumstances, on the context and on the definition of usefulness. In a classic Marxist reading, a good hunting dog has use value for a hunter but would not hold use value for a non-hunter, even though it most certainly could hold other types of value or prove useful in other ways. However, the primacy of use value is striking in almost all these texts. Emotional attachment, while certainly in existence, was not yet a stand-alone value. Training was not about bonding, though it was a legitimate side effect. When Foucault, in *The History of Sexuality, Vol. I* writes of the two poles of biopolitics, he notes, speaking about the effects of biopolitical practices on the human body, "the parallel increase of its usefulness and its docility" (Foucault 1978, 139). Docility and usefulness are inseparable: the purpose of docility is usefulness. It would not make sense to create docility outside of a framework of usefulness.

Secondly, and as a close corollary to the first point, looking upon these almost fifty years as the "dark ages" of dog training makes this dynamic period seem static and obscures the plurality of voices within the dog training community. The first and most important narrative that becomes undiscernible when the period is treated as a homogenous whole is the almost revolutionary shift that takes place in the 1930s when obedience competitions take off. This chapter follows the tradition in which the dog is viewed as a representative of "wilderness" and "nature," while the next one, even though it overlaps in time, tells a different story: that of the dog as an obedient citizen. While the domestic animal always straddles the nature/culture divide, when the century opens, the dog is more "natural" (more like a wolf) and in its fourth decade the pet dog becomes more "civilized" again. This change can be traced through the debate on the training of personal protection dogs. By the time of the publication of Saunders's book, the debate was over, and it was generally acknowledged that the *American* pet should not be taught to bite. The major shift described above has a crucial gender dimension as training slowly begins to be seen as a feminine activity. Even though Konrad Most, Fred Kollett and Josef Weber (with their German shepherds) serve as the icons and anchors of this period, the development of the sport of dog obedience can be credited to two women: Helen Whitehouse Walker and Blanche Saunders (with their poodles).

1 Was Most Modern?

German Colonel Konrad Most's book was published in English only in 1954, which also happened to be the year of the Colonel's death. The first American edition came out in 1955. Yet, Most is routinely mentioned as a significant influence on American trainers and is called the founder of modern dog training by multiple academic English-language sources aimed at the animal behavior community (Burch and Pickel 1990; Burch and Bailey 1999; Lindsay 2001; Pręgowski 2015; Reisner 2016), in addition to dozens of non-academic popular articles on dog training published in the dog press and online. The conundrum of Most's alleged influence is usually resolved through emphasizing the prominence of Most's former students who immigrated to the United States and established training schools. Yet, this explanation still sounds incomplete: what was it really about Most that makes him a household name among American dog trainers?

The word *modern* in the label "father of modern dog training" (Lindsay 2001, 10) deserves a closer look. Most has been described as modern for showing "an understanding of the key elements of operant conditioning including primary and secondary reinforcement, extinction, shaping, fading, chaining and negative conditioning (punishment)" (Burch and Pickel 1990, 263). In other words, he did not bridge the gap between science and its practical applications, but his practical applications predated much of the scientific research on animal learning. However, so did, for example, S.T. Hammond's book from 1882, which also happened to be written in English and published in the US. Furthermore, having at their disposal the same practically acquired knowledge of animal learning – later described using scientific vocabulary by Pavlov, Thorndike and Skinner – Hammond and Most decide to use it in diametrically opposing ways. Why does the contemporary dog training community single out Most as such a major influence? From the perspective of methods used, Hammond certainly seems the more likely figure to be celebrated by reinforcement-based dog trainers as an early pioneer. Yet, this is not the case. Additionally, in the early twentieth century it was not unheard of to find writers who were influenced by the most up-to-date research on animal behavior and cognition. For example, H.W. Carlton's book on the training of spaniels, published in London in 1915 and in the US in 1922, includes a thorough summary of American psychologist Edward Thorndike's work on animal learning and of the research by British ethologist C. Lloyd Morgan.[1] Most's practical understanding of key elements of operant conditioning – while exceptional – was not unique. At least not to the extent that it warrants the label "father of modern dog training."

Most's modernity should rather be diagnosed in his association with the institutionalization of training and the embrace of dogs by the military and the police in the first decade of the twentieth century. In other words, Most – unlike Hammond or Carlton – is associated with the development of the classic disciplinary institution: the dog training school and with inventing techniques best suited to the function dog training plays as a technology of discipline. Clearly, Most was modern in the sense in which disciplinary power is modern. It is not coincidental that the emergence of canine disciplinary institutions correlates in time with the turn towards substitute activities for herding and guarding dogs in an era of increased urbanization. The classic canine activities of herding and hunting were not taught in state-sponsored, institutionalized group classes. Modern training is thus about finding new jobs for displaced dogs and about harnessing and utilizing dogs' potential not just for their individual owners but for the state. Such a connection creates a lineage – a pedigree, so to speak – in which contemporary pet dog training descends from the training of military and police dogs, bypassing all the other possibilities (circus dog training, hunting dog training, herding dog training, etc.), though, as it develops into an activity for pet dogs, attack work is dropped from the repertoire of exercises. This certainly makes sense from a Foucauldian perspective: singling out Most's state-sponsored breeding and training program as a precursor of the pet dog training class traces a genealogy of disciplinary power. This lineage also has empirical grounding: military methods such as those developed by Most indeed trickled down to the general pet-owning population and influenced American pet dog trainers in the early twentieth century probably more than the methods for training hunting dogs.

Indeed, even the exercises most strongly associated with dog training – heeling, staying, coming when called, changing positions (sit, down, stand) and retrieving – are formalized in the first decades of the twentieth century. Heeling, for example, did appear as an exercise in hunting dog manuals from the second half of the nineteenth century but heel position was understood as following behind the hunter's feet (literally, "at the heels"). It is with the entry of dogs into the uniformed professions that heel position becomes defined

1 It should also be added that Carlton's methods are mostly reward-based. Carlton uses food rewards for teaching multiple exercises, advocates the training of puppies. He speaks of training as "perhaps too harsh a word to apply in a process that consists of an easily graduated series of lessons given without a trace of asperity" (5).

as the dog's shoulder being in line with the handler's body, as a classic disciplinary technique facilitating close surveillance: "the chest of the moving dog should be practically in line with the left knee of its trainer so that the latter may keep the animal constantly in view" (Most 43). From a practical purpose, this position allowed the handler to supervise the dog more closely while keeping his right hand free for actions other than holding the dog's leash. A 1931 article from the London *Times* reported that: "Trainers of Alsatians in Germany adopted a new position for 'heel' that has since been accepted by our own with advantage. Instead of 'brushing his master's spurs,' as Russian sportsmen used to term it, a dog is expected to follow closely at the left side, where he is under observation without one having to turn the head to ascertain if he is there" (*Times* 1931, 15). The very invention of the heel position as defined in Most's book is a step towards further refinement of disciplinary techniques: constant supervision is, after all, a staple of discipline.

Still, the main problem with the argument I have just sketched – that is of Most's significance as a precursor of discipline – is Colonel Konrad Most's role in the entire process. Most's book was not unique: it was not the only manual on dog training published in early twentieth-century Europe. Chris Pearson, writing about police dog training in early twentieth-century Paris, lists similar books published in Belgium and in France at roughly the same time (Pearson 2016). Neither was the institution headed by Most – since 1912 he served as director of the State Breeding and Training Establishment for Police Dogs in Berlin – the only one of its kind in Europe. In fact, Belgium is usually listed as the main innovator in police dog training, establishing a school in Ghent in 1899 (Chapman 1990, 10); while the use of dogs to aid policemen spread to other Western European countries at more or less the same time: the first decade of the twentieth century (Pearson 470–472).

Furthermore, there existed publications on the training of "police dogs"[2] available to the American reader decades before Most's book was translated: Robert Gersbach and T.F. Jager's *The Police Dog in Word and Picture* was published in New York in 1910 and was heavily advertised in popular dog magazines of the time. The training section of Max von Stephanitz's book *Der Schaferhund in Wort und Bild* was translated by Anne Tracy as *Schooling and Training the German Shepherd Dog* (1917) and David Brockwell's *The Police Dog* (1924) also included a translation of the same section from von Stephanitz's

2 The term police dogs is placed in quotation marks because these books were aimed for a general readership not for police dog handlers, as implied by the types of the publishing houses issuing them, the formats of publication, the marketing strategies used to promote them and, finally, the contents of these books.

book, while the complete translation of this work was published in America in 1925. A book titled *Training the Police Dog*, by German-American trainer and breeder of German shepherds Fred Kollett, was published in 1924 in popular format by Judy Publishing in Chicago.[3] In short, the American public had plenty of opportunity to read about the type of training advocated by Most well before his book was translated into English. While it is true that Most's German students who immigrated to the United States, specifically Josef Weber, Hans Tossutti and Carl Spitz – who all operated training schools and published their own books in the 1930s or 1940s – were influential in popularizing his methods, Most's own role as "the father of modern dog training" is more symbolic than literal for the English-speaking world. However, actual human beings are turned into symbols because they best epitomize the zeitgeist of the period. There was something in the figure of Most – in his ideas and in his book – that aided his canonization, even if this canonization happened *post factum* and results from a particular type of emplotment. Most is almost shorthand for a certain approach: one that traces back to Germany and that pushes for heavy-handed training methods falling back upon a philosophy of canine usefulness to mankind to justify compulsion; and one in which the dog is a threatening, almost wild animal. At the same time, Most's name is associated with the codification of the disciplinary regime as it relates to dogs: with the establishment of the dog training school. While this was not yet a school for every dog, the police dog training school served as a model on which pet dog training schools would later be based. Most's book, while not the only one proposing a certain training philosophy, is the most comprehensive and very well written. It reads almost as a work of good fiction. Finally, Most's ideas about the human-canine relationship in training are exemplary of the changes in the perception of the man/"nature" relationship that took place, more or less, at the turn of the century.

2 Police Dogs

Americans' fascination with German shepherds dates back to the second decade of the twentieth century, which means that it shortly follows the breed's codification by Max von Stephanitz in Germany and, in a cultural perspective, is also closely tied to the revaluation of the meanings of "the wild" that

3 Fred Kollett's methods, exercises and even the comparisons to children's learning that he uses are so strikingly similar to Most's methods that an accusation of plagiarism could possibly be made.

took place in turn-of-the-century United States. In fact, the German shepherd played a double symbolic role: the dog was both a link to nature ("the wolf") and the epitome of wild justice, a kind of cowboy figure for the "urban jungle," a term popularized in turn-of-the-century literature about the city. In popular understanding, the German shepherd dog, often referred to as the wolfdog, was a direct descendant of the wolf, the breed most closely related to the wild *canis lupus*, though improved through selective breeding: a kind of superwolf.[4] On a symbolic level, owning a German shepherd dog brought one closer to "nature" than the ownership of breeds that had been more altered by human intervention (particularly by misguided intervention seen as leading to over-breeding). Interacting with wild nature was increasingly seen as something of a vaccine against the degeneration of modern civilization, as evidenced by the growing popularity of outdoor activities, such as camping, big game hunting, the emergence of boy scouting, etc. (Haraway 1984; Bederman 2008). The fear of racial degeneration was, not surprisingly, connected with eugenic discourses that were coming to prominence in the first decades of the twentieth century, and which had not yet been corrupted by their later association with the atrocities of Nazi Germany. As Aaron Skabelund notes in "Breeding Racism: The Imperial Battlefields of the 'German' Shepherd Dog": "the revaluation of local dogs and their identification with the nation-state emerged with the rise of nationalism and the growth of an urbanized, bourgeois middle class, whose members were nostalgic for the countryside" (Skabelund 2008, 356). While it may seem strange that it was the German shepherd dog that served as the necessary link in the connection of urban dwellers to "nature" in the United States, Skabelund, though analyzing the Japanese context, posits that the breed's identification with Germany was superseded by its association with the dog's role as the enforcer of the law: in the case of Japan, an enforcer of colonial rule. The dogs became similarly popular and significant in the United Kingdom, where they were renamed Alsatians after World War I to sever their ties with despised Germany. I have written elsewhere about a similar renaming strategy used in post-World-War-II Poland (Pręgowski and Włodarczyk 2016).

The strategy of renaming operated somewhat differently in the United States. During World War I the American breed club for German shepherds was renamed as The Shepherd Dog Club of America but very quickly "German" shepherds, which truly were imported from the most renowned kennels

4 It must be added that while Von Stephanitz was absolutely fascinated with the dog's wolf heritage, he did not claim that the German shepherd was a direct descendant of the wolf.

in Germany, became known as "police dogs," in blatant disregard of the fact that the first actual dogs employed by the police in the US at the time were not German shepherds (Figure 3).[5] However, this strategy, in addition to disidentifying the breed from its association with America's enemy, helped the German shepherd be perceived as the enforcer of the rule of law, even if the dog was in private hands. As Skabelund argues, writing about the Japanese context: "army dogs (…) were often used to magnify the martial power and authority of their master – both military and civilian" (Skabelund 2008, 367). This was clearly also the case in the United States in the 1910s and 1920s. It should also be added that demand for German shepherds skyrocketed after the Rin Tin Tin series of movies was filmed throughout the 1920s (Orlean 2011). In popular German shepherd history, the 1920s in the United States are referred to as the era of commercialization: overbreeding accompanied the breed's rise in popularity (Dodge and Rine 1956, 6–7).

The German registry for purebred German shepherds – known under the acronym SV, standing for Verein fur deutsche Schaferhunder, and established in 1899 – began implementing a system of breeding selection that included a component testing the dog's "intelligence" in addition to his conformation to a written breed standard. These "intelligence tests" were in fact competitions that evaluated the dog's training, although here training and breeding went hand in hand: trainability was firmly judged to be an effect of breeding. The training exercises included protection work: defending the owner against a mock attack and seeking out a mock criminal. The identification of the "German shepherd" with the "police dog" in the 1910s and 1920s has resulted in confusion related to the history of actual police dogs, the history of German shepherds in the US and the history of pet dog training. David Brockwell's 1924 book *The Police Dog* is in reality something of a "How to Raise and Train Your German Shepherd" guide.

5 Chapman's *Police Dogs in North America* lists Belgian sheepdogs (imported from Ghent, Belgium) as the first official police dogs in New York City. However, when one examines pictures of police dogs from Ghent from the period, it is clear that they do not look like contemporary purebred Belgian sheepdogs. It is thus likely that the designation "Belgian sheepdogs" used in reference to the Brooklyn police squad from 1907 refers simply to dogs brought over from Belgium. In any case, they would not have been "German shepherds." Chapman also lists St. Bernards as the most likely candidates for America's first police dogs. It is clear that the popularity of dogs matching the "German shepherd" phenotype in the US actually postdates their popular designation as police dogs. It is an interesting case of real-life use succumbing to expectations created by the aggressive marketing policies of German shepherd breeders. The increased demand for pet German shepherds in the late 1920s and 1930s, resulting from the popularity *of Rin Tin Tin* movies, additionally provided a ready supply of domestic-bred dogs that could be used by the police.

FIGURE 3. Photographs of early police dogs in New York City suggest that most of them were not German shepherds. Bain News Service. 1912.
COURTESY OF LIBRARY OF CONGRESS, LC-DIG-GGBAIN-00004.

Some of the other books with the words "police dog" in the title, for example Kollett's *Training the Police Dog*, were actually general training books based on preparation for the SV's so-called Schutzhund exams and were addressed to the pet dog owner, as evidenced by chapters on teaching a dog house manners and preparing him for conformation dog shows. Trials inspired by those organized by the SV in Germany began to be organized in the US in the early 1910s. A 1913 article from the *The New York Times* describes a trial for "police dogs" organized in Van Cortlandt Park in Brooklyn, New York City, attended by an audience of 5,000 spectators. The exercises in the trials included so-called man work, that is attack work and the apprehension of mock criminals in protective clothing but also exercises in obedience and feats demonstrating the dog's agility and physical prowess (*New York Times* 1913, 5). In *The Story of Dog Obedience* Blanche Saunders includes a letter from Hermann F. W. Steinn, who recalls trials organized in the 1910s on Staten Island and championships for Police Dogs organized in the 1920s (Saunders 1974, 31). It is clear that the dogs taking part in these "police dog" trials were owned by private individuals (and/or private breeding kennels), not by actual police officers. In fact, one of the two equal winners of the 1913 trial was a dog owned by Miss Anne Tracy, one of the founders of the German Shepherd Dog Club of America.[6] From a contemporary perspective these "police dog" trials may seem to gravitate between the ridiculous and the brilliantly gender-bending (when one considers Anne Tracy, a wealthy New York socialite in the role of the courageous police officer), but they fulfilled a certain symbolic purpose: they imbued the keeping of (basically pet) German shepherds in the city with a deeper ethical dimension through the utilitarian link. They posited an alternative function for dogs in the city; one which broke away from the increasingly corrupt image of the "lap dog," associated with uselessness, owner's self-indulgence and, in terms of training, the performance of tricks for sheer entertainment of the owner's guests.

3 Most's Masculine Methods

While the organization of the process of training proposed by Most and his followers ushers in the disciplinary regime, the methods proposed by the trainers are not always exemplary of discipline: they rely too much on spectacle, confrontation, on unnecessary exertion of the trainer's energy. This is not yet the efficiency one associated with classic discipline. There is an excess in Most that is incompatible with discipline's guiding principles. If, as Skabelund suggests, the "German" shepherd functioned as a generic representative of colonial rule, the dog first had to be turned into a tool of imperialism through a process which enacted the very dynamics of colonial rule that the dog was later supposed to embody. In order to become a tool of power, he first had to be dominated and subjugated in much the same way in which he was then expected to treat the "enemies" of the state: through physical confrontation. While Most's influence on American training is not as direct as it is sometimes made out to be, both his methods and his philosophy are certainly representative of this shift towards perceived canine usefulness in the service of the nation-state. However, it is difficult to analyze his methods without noticing how embedded they are in the broader understanding of the man/nature relationship and the redefinition of white masculinity.

In her influential book *Manliness and Civilization* (2008), Gail Bederman identifies a crisis of the broadly Victorian notion of manliness, which prioritized self-control and self-sacrifice, and the gradual emergence of a more virile masculinity which emphasized "aggressiveness, physical force and

6 It is unclear whether Anne Tracy showed the dog herself or if she was simply the dog's owner.

male sexuality" (17–18). Bederman places these changes at the turn of the nineteenth and twentieth centuries. Overcivilization begins to be seen as a problem to be overcome through vigorous exercise, hunting, fishing and similar activities. However, interacting with a dog could prove a substitute of the above-mentioned activities for the male urban dweller as long as the activity was structured through ritualized interactions that strengthened the trainer's masculinity as they reenacted the drama of man's conquest of nature. This is exactly what the German trainer's book offers. It reads much like a work of naturalist fiction: a prescriptive version of Jack London's *White Fang* for the urban dweller, though maybe a cross between *White Fang* and *The Call of the Wild* because the dog ends up domesticated as a result of submitting to the human's superior willpower and strength, not as a result of the human's benevolence and the dog's implied gratitude. Most thus writes about the human-canine relationship as one in need of constant verification:

> As in a pack of dogs, the order of hierarchy in a man and dog combination can only be established by physical force – that is by an actual struggle in which the man is instantly victorious. Such a result can only be brought about by convincing the dog of the absolute physical superiority of the man. Otherwise the dog will lead and the man follow. If a dog shows the slightest sign of rebellion against his trainer or leader, the physical superiority of the man as leader of the pack must be given instant expression in the most unmistakable manner. (Most 1955, 25)

The logic governing the above quote is fundamentally different from the approach described in chapter 1, where practical matters prevailed when compulsion was applied. Despite the rhetoric of gentleness, nineteenth-century trainers turned to the whip whenever they could not come up with other means of eliciting a desired behavior. If eliciting a behavior seemed simple without the use of force (such as in the case of teaching a dog to stand up on his hind legs), they used food lures. The style of training represented by Most is vehemently opposed to food rewards. Here, training is not civilizing; it is taming. The dog, for the first time in the period under discussion in this book, becomes a wild beast; an analogy that would have been inconceivable in the nineteenth-century mindset. The seriousness of the undertaking of training justifies the use of compulsion as something of a pre-emptive tool: the establishment of a hierarchical human-canine relationship through the use of force is a pre-requirement of training. At the same time, the use of compulsion is mitigated through a vocabulary of – ironically – humane treatment of the dog (75–76). Because reliable performance is required in a working dog, it is said to be more humane to use harsh compulsion (as opposed to mild compulsion) to prevent the necessity of repeating it too often: "In the first place, too mild compulsion may render a dog unreliable in a number of exercises when the real services of a working dog are required" (25).

Still, the paradox at the root of this logic is quite blatant. Most speaks of the goal of training as "permanent and unconditional surrender of the dog" (27); yet, the entire program of training and proofing is based on lack of permanence: on eliciting repeated confrontations that require physical resolution. The hierarchy needs to be verified over and over again. It is significant in relation to the discussion of the new hegemonic masculinity that emerges in the early twentieth century that Most speaks of the pack hierarchy as never firmly established but as something that must be proven on a regular basis. The trainer should expect challenges from the dog and react to them accordingly, that is with the necessary physical force: "For each time the dog finds that he is not instantly mastered, the canine competitive instinct will increase and his submissive instinct will weaken" (26). This idea has haunted dog training for many decades – and still continues to do so in the methods of, among others, Cesar Millan – resulting in training programs that advocated constant watchfulness for signs of the dog's rebellion against the owner. It is telling that within masculinity studies, masculinity is often defined as a notion that needs to be repeatedly verified through ritualized displays (Kimmel 2001, 276). Here, in keeping with this idea, the human-canine relationship is a site that provides endless opportunities for such ritualized performances of masculinity. The relation is both a performance of human mastery of the dog but is also very much performative in Judith Butler's understanding of the word: that is the hierarchical relationship is "real only to the extent that it is performed" (J. Butler 1990, 2). It is in the ritual performance of submission, in the repetition of the "primal scene," that of man's conquest of the "wild animal," that the relationship of superior man and inferior dog is continually verified. Indeed, the kind of training methodology represented by Most relies on almost constant "proofing": on setting a dog up so that he will fail in the correct performance of the task, in order that a physical correction can be administered. The endless repetition of the dog's inability to achieve the goal set forth for him is a constitutive element of training, NOT its failure: "During training the trainer should continually give the dog fresh opportunities to leave his left side so that he may apply the necessary correction. It is necessary for him to leave and to experience discomfort as a result, and for this purpose, as already emphasized, regular loosening of the lead is essential" (Most 1955, 47). In the US, Robert Gersbach and Jager write in 1910: "Work him till he fails to obey, then force him to obey, as this is the only way for him to subordinate his desires to those of his master" (Gersbach and Jager 1910, 98).

Most is also apparently the first trainer who uses the term "pack hierarchy" to refer to the human-canine relationship. After Most, the idea of the pack – and specifically of pack order – was a notion that shaped dog training for decades. Trainers like Cesar Millan continue to posit that the human has to become the "alpha wolf" in his relationship with the dog. However, it is important to notice that Most does not speak of the "alpha wolf" and does not even use the wolf analogy in his writings. For him, the human-canine hierarchy is based on the hierarchy in a pack of wild dogs. While recent research, specifically Raymond and Lorna Coppinger's work on feral dogs (Coppinger and Coppinger 2016), proves Most's understanding of wild dogs wrong, Most is often credited with introducing the erroneous wolf model into dog training. This is not the case and most likely results from a conflation of popular ideas about the German shepherd's wolf heritage and Most's notion of the pack order, or possibly, from an internalization of how these methods fit into the understanding of the man-nature relationship in the early twentieth century. Because Most functions as the Jack London of dog training, he automatically becomes associated with the wolf analogy. However compatible this analogy seems with Most's views, he is not its author. In reality, the wolf pack approach, with its assumption that the trainer must establish himself as "the alpha wolf" in his relationship with his dog using techniques such as "the alpha roll," was codified much later, at least in the English-speaking world. It is the Monks of New Skete with their – somewhat ironically titled – 1978 book *How to Be Your Dog's Best Friend* who first wrote in detail about how understanding wolf behavior should aid the trainer in understanding his dog.[7] It is highly interesting in itself that contemporary trainers tend to view the "alpha wolf" model, which they vehemently oppose, as "ancient history," even though it seems to date back only to 1978. Nineteenth-century writers did not dwell on issues of pack hierarchy.

It must yet again be emphasized that Most's confrontational attitude coupled with the turn to physical force for resolving conflict situations was not unique. It was representative of the general thrust of training methodology for the training of police and army dogs in the first decades of the twentieth century, also in the US. Fred Kollett wrote in 1924: "Should the dog ever show any

---

7 The popularity of the wolf analogy in dog training was also aided by the publication and subsequent popularity of Konrad Lorenz's *Man Meets Dog* (German: 1949, first American edition: 1953), which argues that certain dogs are descended from wolves (while others from jackals) and that the wolf dogs treat their masters as pack leaders. Towards the end of his life Lorenz refuted these hypotheses (Coppinger 2001, 36). Additionally, studies of dominance in wolves carried out in the 1940s by Rudolph Schenkel (and which have since been disproved) probably influenced dog training. However, the wolf analogy does not come into popular use in dog training vocabulary prior to the 1970s.

signs of attacking his trainer or growling at him, or trying to bite, here you must use force and whipping is necessary, and you must not give up until the dog lays down and whines, as it is a case to show who is master" (Kollett 1924, 10). Theo Jager thus described preliminary heelwork in 1917:

> Go to the kennel and apply the spiked gag collar, to which the leash is attached. With the exception of the command remain perfectly silent during this entire exercise. [. ...] Walk slow and command "To-heel" while starting. Repeat this at various times, whenever you note a drag on the leash. If the raw dog whines or throws himself when the spikes hurt him, or when he pulls back, stop momentarily, but continue to pull on the leash in continuing your walk. Pay no attention to dog and start slowly, even if you are compelled to drag the dog on his haunches over the ground. Exert the slow and steady pull on leash at all times, and the dog will eventually give in to your will and follow you, a dejected looking and brokenspirited animal, that has comprehended his first exercise. (Jager 1917, 63)

The "brokenspirited animal" is not an unfortunate byproduct of a crude method of heeling training but a desired effect, if one thinks of training as the confrontation between a wild animal and a human being, intent on performing the superiority of the human. Many contemporary trainers wonder about why "in the olden days" trainers were not able to elicit joyful performances from their dogs: why dogs lagged while heeling, why they did not look at their owners during obedience exercises. The reason was not simply lack of trainers' skill but a different understanding of the purpose of training. The performance these "old trainers" were staging was a different type of a performance: not one of joyful collaboration but of canine submission to human authority. If we go back to Jane Desmond's idea of animal performances as a restaging of what is considered the dominant representation of "nature" at a given point in time (1999), it becomes clear that this is a performance of human domination over nature; and not just of benevolent dominion but of violent conquest.

If we are to use a more Foucauldian vocabulary to conceptualize this phenomenon, it can serve as a striking example of Foucault's understanding of the power-knowledge relationship; that is of power being based on knowledge but, on the other hand, also reproducing knowledge by shaping it. In "Truth and Power" Foucault speaks of "seeing historically how effects of truth are produced within discourses which in themselves are neither true nor false" (Foucault 1984c, 60). Here, the knowledge of the dog as wild beast shapes the power relationship in an obvious way: the wild animal needs to be dominated to render it unthreatening. At the same time, the very brutality of the physical confrontation strengthens the knowledge of the dog as a beast. This "truth effect" becomes particularly glaring when placed in a historical context: just thirty years earlier dogs had been "sagacious creatures" and no one conceptualized training practices in terms of exerting physical power over the dog.

When viewed from this perspective, such performances were successful. An eagerness for work, not necessarily exhibited through what are nowadays considered the behavioral signs of canine pleasure in human-canine interaction (wagging tail, maintaining eye contact) but through prompt responsiveness to commands, could be – in the eyes of Most – obtained through proper application of compulsion. If compulsion was stopped the moment the required response began – that is if, in B.F. Skinner's behaviorist language, negative reinforcement was applied skillfully – a prompt, if not overtly joyful, performance could be obtained. Most writes: "The result of this liberation from the pressure of compulsion is that the dog quickly learns how to escape his disagreeable experience (…). This causes him to develop an amazing zest for his work" (Most 1955, 23). In fact, negative reinforcement – which is what Most is referring to in the above quotation – turned into the primary strategy for ensuring the performance of desirable behaviors with speed. A very famous training method that left its mark on dog training for years was the so-called forced retrieve, or ear-pinch retrieve, actually developed (or at least codified in writing) by hunting dog trainers in the nineteenth century. In this method, the dog's ear is pinched with the trainer's fingers until the dog opens his mouth to squeal out in pain. Once the dog's mouth is open, a dumbbell is promptly placed inside and the pain ceases. After following protocol, the dog reacts to just a touch of his ear by springing to his feet and running full-speed to pick up the dumbbell. Most himself did not use the ear pinch but did write about the use of negative reinforcement in teaching the retrieve: he gagged the dog with a prong collar until the dog opened his mouth (127–128). To be fair, it must be added that the technique was reserved for difficult cases.

4 Nietzsche Goes to the Dogs

Compulsion, according to Most, needs to be applied promptly and with sufficient force: this is actually something contemporary behavior analysts agree on with Most, although they invariably list the negative side effects of relying on compulsion. It most certainly is true that punishment works only when it is severe enough and when it is not distanced in time from the behavior which warrants its use. Most, however, embeds the use of compulsion in his signature theory of canine instincts, among which he includes: the pack instinct, chase instinct, defensive instinct, instinct for freedom, competitive instinct and submissive instinct. In the nineteenth century, training (particularly of pet dogs) did not dwell on a dog's instincts and, as chapter 2 argues, even in the case of hunting dogs, it was control over instinct that was prized the most highly. Instinct was seen as opposed to intelligence and training was the province of extending (and proving) a dog's intelligence. In the case of hunting dogs, instinct was needed but pure instinct was never something positive: the purpose of training was to establish control over instinct. This was achieved mostly through proper breeding. If that failed, a beating could possibly still work, though the case was almost hopeless. Beginning with Most, training starts to be seen as the manipulation of instincts in a way that suits the trainer's needs, but not simply the suppression of instinct. While according to Most, the dog's submissive instinct aids the dog's learning, his competitive instinct must be weakened in order for learning to take place. Similarly, as a prerequisite for training, the dog's instinct for freedom needs to be quelled. While the above examples are nothing short of ludicrous from the perspective of contemporary knowledge about animal learning and dog behavior, Most's application of what he called "the chase instinct," that is using elements of the dog's predatory behaviors in order to facilitate training, has played a role in the development of today's positive training methods, used, for example, in agility or flyball training.

A similar revaluation of the role of instinct takes place in the work of Friedrich Nietzsche. While I have not conclusively established Nietzsche's direct influence on Konrad Most, it is clear that so-called force-based training methods can be – and have been – read as an application of certain Nietzschean concepts. For example, an exegesis of Nietzsche's understanding of the notion of responsibility has been carried out by Randall Havas through an analysis of responsibility in the work of dog trainer William Koehler, a successor to Konrad Most's thought on training (Havas 2012). For Havas, dog trainer William Koehler's writings are basically a practical application of Nietzsche's ideas. Certainly, both Nietzsche's philosophy and Most's training methodology can be read as a reaction to Victorian culture's prioritizing of self-control over impulsive behavior. In fact, Nietzsche's entire revaluation of Christian morality in *The Genealogy of Morals* (1921 [1887]) has often been read as arising from a suffocating Victorian culture that made Nietzsche "overreact" to Christianity's valorization of meekness and gentleness (Loeb 2017). Nietzsche's critique of Christian virtue is steeped in his repulsion toward the obedience to authority that morality requires:

> Morality is preceded by *compulsion*, indeed it is for a time itself still compulsion, to which one accommodates oneself for the avoidance of what one regards as unpleasurable. Later it becomes custom, later still voluntary obedience, finally almost instinct: then, like all that has for a long time been habitual and natural, it is associated with pleasure – and is now called *virtue*. (Nietzsche 1921, 136)

It would seem that the process required for the development of the type of morality Nietzsche despises is exactly analogous to the process of training, as conceptualized by Most. Logically, the (trained) dog should not be held in high regard by Nietzsche. And indeed, that was the case. Nietzsche prioritized wildness over domesticity: the wolf over the dog. He admired the lion's "will to power" and despised "herdlike" behavior. However, at the same time Nietzsche is fascinated with the simplicity of animal morality, which he does not see as inferior to that of humans. On the contrary, as Claire Rasmussen sums up Nietzsche's thought from *The Genealogy of Morals*, "reliance on their drives preempts animals from developing elaborate codes of morality that would prevent natural predators from consuming their natural prey. In short, animals lack the kind of self-limiting tendencies in humans that Nietzsche sees as causing weakness" (Rasmussen 2011, 107). Engaging in an interspecies relationship where a member of one species clearly imposes his will on a member of the other species could, potentially, be read within a Nietzschean framework as a step towards reaching the ideal of *Übermensch*: the new man, unhindered by Christian morality, life-affirming and creative. This potential Nietzschean reading of "the primal scene" of dog training – the physical enforcement of the dog's submission – complements my previous reading of this scene as the re-enactment of man's conquest of nature: the *Übermensch* basks in his masculine strength.

Additionally, within this Nietzschean framework, a training methodology that recognizes the dog's animality and does not stifle it – as the nineteenth century handbooks did – carries more positive value not only for the human but also for the dog. Cary Wolfe, in his reading of Nietzsche's *Zarathustra* in *Before the Law*, argues that the cardinal sin for Nietzsche was "the deadening and diminishing of 'animality' itself, in all its vitality, creativity and multiplicity" (Wolfe 2013, 41). While Wolfe uses this argument to devise a critique of factory farming, which creates conditions in which animals' vitality is diminished, it is easy to see how the Nietzschean valuation of animality can be used to argue for a training methodology that remains – or claims to remain – closer to the dog's "animal nature." Within a Nietzschean framework, an animal training methodology that prioritizes instinct over self-control, as Most's methodology does, can be read as more dog-friendly because it emulates rules of nature. It does not stifle and suppress what is seen as the defining feature of animality; it builds on it. If dogs resolve problems through force, it is only natural that humans need to apply force in interactions with dogs. It is also makes sense to employ dogs' natural proclivity for aggression in the service of humans. The problem that contemporary ethology would most certainly have with Nietzsche is that his animal metaphors do not correspond with the specificity of actual canine social life. Nietzsche sees physical struggle and conflict everywhere. Contemporary ethology places a lot of emphasis on the avoidance of conflict among animals; not to mention the increasing number of studies that locate the origins of human morality in an animal morality, such as Frans de Waal's work on bonobos (1996; 2010), and that see dogs as capable of understanding concepts such as fairness (Hare and Woods 2013).

From a Nietzschean dog trainer's perspective, what is to be admired in the dog is precisely the instinctiveness of his behavior; not his "sagacity." Nature, not culture. Most's distinction between intelligence and trainability can also be read through a Nietzschean framework. However, this reading makes sense only if trainability is seen as akin to ease of submission. A trainable dog is one who is easily dominated; an intelligent dog is one who does not submit. Intelligence, argues Most, has nothing to do with trainability: "It should be stressed that the number of repetitions does not depend exclusively on the 'cleverness' or 'stupidity' of an individual dog, but that the condition and strength of the animal's instincts also affect the situation" (Most 29). In another move that ties him to Nietzsche, Most saw the imposition of a human notion of morality – so key in the nineteenth-century view – to be not just an undue burden on the dog but also as a path to mistreating the dog: "This system avoids the mistake of endowing dogs with human understanding and morals – a mistake which renders any real partnership between man and dog so difficult" (Most xii). In acknowledging the dog's lack of higher reasoning powers, Most and his followers unanimously see themselves as doing the dog a favor: taking the burden of morality off the animal's shoulders (Most 75–76). Most's student Josef Weber writes: "Just because he is only a dog and has no highly developed reasoning power but possesses only instinct, we, to achieve a better mutual relationship, must remain man and seek to appreciate and understand the dog from the standpoint of his own lower level of development" (J. Weber 1938, 15). Meanwhile, Fred Kollett states bluntly: "The dog is unable to think, therefore, he must have a human leader" (Kollett 1926, 7). Even if Most values canine instincts, he does see the need for curtailing some of them: he speaks of the need to quell the dog's instinct for freedom. Training is necessary not because it elevates the dog's mind but because it transforms the body into docility. Chris

Pearson's analysis of police dog training in early twentieth-century Paris also suggests that training was seen as having transformative value, which proves that these beliefs were not unique to the English and German-speaking worlds. As Pearson writes about the French trainers: "police dogs would emerge through trainers suppressing some canine instincts and nurturing others. Without their guidance, trainers feared that their dogs' aggressive or wayward instincts would dominate" (Pearson 2016, 466).

Pearson also notices the importance of the city as the space of human-canine interactions in early police dog training. In his work on police dogs in Paris, Pearson traces how the city increasingly began to be associated with violent crime perpetrated by so-called Apaches. The dogs were "refashioned by police officials and dog aficionados as allies in the fight against the criminalization of public space" (469). The association of the name of a Native American tribe with violent criminals is, of course, not a literal designation: there were no actual Native American immigrants in Paris. It is, however, symptomatic of the wider anxiety related to the mixing of classes and nationalities sweeping the early twentieth-century metropolis, a phobia that the United States was not immune to, although the culprits were differently identified. Bederman writes of the city becoming a space closely associated with marginalized groups, particularly immigrants and black people migrating from the South after the Civil War. Building on Bederman's ideas, Anna Pochmara writes: "In the dominant mythology, they [immigrants] were perceived as less virile races, positing the threat of cultural degeneration" (Pochmara 2009, 134).

The figure of the trained German shepherd police dog ties the loose ends of some of the tensions that define the opening of the twentieth century: nature/culture, wilderness/city, progress/degeneration. In this context, the police dog served a double role: he was at once a wild other (a potential "Apache" himself) but he could also be turned into a tool in the fight against the "Apaches" if trained properly and constantly monitored. The recognition of the dog as an animal other was a logical necessity in a framework in which the human/canine relationship was fundamentally a recreation of man's conquest of nature. The dog had to be that wild other to be conquered and to serve the purpose of proving the strength of the urban man. As Pochmara writes in her study of race and early twentieth-century masculinity: "Victorian manliness draws privilege from the position of patronage over uncivilized non-white races, whereas modern masculinity appropriates the imaginary primitive vitality of other races to revitalize itself" (Pochmara 2011, 10). Here, the confrontation is not between different races but different species. It is the dog's supposed wildness, much like the savage's primitiveness, that serves as the source of primitive vitality. But the dog is not just the primitive other. His role is more complicated: through this confrontational interaction resulting in the human's victory, the dog also becomes a tool in the police man's hands: a tool in the fight for, what Donna Haraway referred to in her essay on Theodore Roosevelt as the "regeneration of a miscellaneous, incoherent, urban public threatened with genetic and social decadence, threatened with the prolific bodies of new immigrants, threatened with the failure of manhood" (Haraway 1984, 23).

5 Should American Dogs Bite?

Most of the textbooks published in the US in the second and third decade of the twentieth century, included guarding and attack exercises. Many of these books included the words "police" or "army" in the title but were, in fact, intended for and purchased by a general pet-owning population. Even though Konrad Most derived his credibility from his work with police dogs, *Training Dogs* was not published as simply a manual for police canine handlers but as a general training book, or at the least was received as such. Most advocated the training of "a mere house dog" for protection work (133). At the time of the publication of Most's manual (1910), the SV had already begun its protection dog program (1901), which included a component of defense work. In fact, Most opens the book with recollections of how he participated in such protection dog trials "many years ago" (xii). Significantly, the trial program seemed to prioritize a certain aesthetics of performance that Most found incompatible with actual police dog work. In fact, he even devoted a section of the book to correcting misconceptions about the work of the real police dog. Specifically, during trials dogs were expected to withstand beatings with a stick that took place during the attack. Most claimed that teaching dogs "stick-sureness" puts them in unnecessary danger and posited that it is much safer and more effective to teach a police dog to avoid the stick: "Unless we provide dogs with armor, we should never train them to be steady under the stick" (167). The tests promoted a certain vision of a courageous dog who does not back out of difficult situations; an idea that was certainly phantasmatic. Avoiding the beating would have been perceived as cowardice on the part of the dog: an interesting case of imposing human-defined cultural standards of evaluation on responses that were supposed to be instinctive. The image of the police dog withstanding beating with a stick can serve as a visual icon of the title of this chapter: the culture of instinct. What was perceived as pure instinct (the dog's unresponsiveness to the blows, evaluated as courage), was in fact the result of a process of training that taught the dog not to fear such blows (thus, of culture).

At a time when anxiety about crime was on the rise, the idea of keeping dogs for self-defense against urban criminals was catching on. As Pearson writes, the Union of Guard and Police Dog Enthusiasts in France "portrayed police dog training as a useful 'recreational sport' suitable for all and an effective form of self-defense fit for a democratic republic" (Pearson 2016, 473). As newspaper accounts of police dog competitions in the United States prove, in the trials organized by the German Shepherd Dog Club of America in the 1910s and 1920s protection work functioned as a crowd pleaser. Even after the (now renamed) Shepherd Dog Club began offering trials that led to the C.S. (or companion shepherd) degree, the exercises included protecting the master without command, guarding objects, being stick proof and gun proof (Kollett 1924, 33). Interestingly, in these 1920s trials the dogs were required to do their heeling exercises while muzzled. It is hard to evaluate whether there was an actual need for the muzzle or if this accessory was yet another element of the cultural aesthetics of the performances, making the dogs appear more dangerous than they actually were. Possibly, this regulation could have been one of the few cases of actual compliance with city ordinances requiring the muzzling of dogs in public. In any case, in Northeastern urban United States a trained dog became synonymous with a dog that could bite an assailant, if required. This definition was, however, short lived: it lasted for no more than two decades.

This state of affairs begins to change in the 1930s with the introduction of the American Kennel Club's program of obedience trials and the popularization of training clubs and training classes. This was a revolutionary change: training of the pet dog was redefined from making a dog bite to ensuring that the dog does not bite. Philip Howell, in his study of dogs in Victorian England, *At Home and Astray* (2015), traces the changes in muzzle and leash laws toward the end of the nineteenth century in Great Britain. Howell notices how the shift from muzzle laws – justified through medical fears: the possibility of the spread of rabies – to leash laws reflects a larger change in attitudes toward dogs in public space. The acceptance of dogs in the cityscape was becoming more common but only under certain conditions: they had to be seen as remaining under "reasonable control" of their owners. The shift towards leash laws also normalized the dog/walker pair, as opposed to the free roaming dog, "as the sign of the respectable and responsible pet-owning citizen" (Howell 2015, 170). Howell theorizes this shift in the management of dogs in urban space as a move from Foucauldian disciplinarity, which he associates with the model of the muzzle, towards governmentality, as represented by the leash. As Howell writes: "the first forwards a more overt kind of control, the medical policing of unsanitary and dangerous animals, the second more obviously involves what Foucault describes as 'the conduct of conduct,' operating through a model of the responsible pet owner rather than directly upon the animals themselves" (151). While Howell's model has its use for speaking about dog training in America, some disclaimers are due first. Most importantly, Howell does not speak of training but of the management of dogs (and, in the case of leash laws, dogs accompanied by humans) in public space. While training does begin to serve the goal of governmentality, that is it becomes subjugated to the major goal of creating docile canine bodies that will not bite the passerby or maim their owner's children, it still does so – at least for a while – using the classic techniques of discipline; an idea that is compatible with Foucault's description of the relationship between discipline and governmentality. This change in the purpose of training is going to influence a change in methods, which will begin to address the dog as a subject capable of self-governmentality, at least to some degree, but this will only come later. At this point, even though the techniques of dog training are changing very gradually – there is no sudden about turn – "protection" exercises are dropped from the repertoire of behaviors taught to an increasing number of dogs. Secondly, the turn towards governmentality in the management of dogs and their people in the United States comes later than in England, in the early 1930s, and is one of the factors explaining the short-lived popularity of the personal protection dog. Thirdly, the turn away from the perception of the dog as a potential aggressor (as represented by the muzzle) to that of a companion (as represented by the leash) is advocated in the United States as distinctly American, as emerging in opposition to the European model.

Even though the Great Depression only strengthened the crisis of masculinity that could be resolved through rigorous physical activities, ownership of a personal protection dog ceased to be seen as an acceptable way of proving one's masculinity among middle-class urban dwellers. Several factors account for this change of heart; some of which are structural and some which may be termed incidental. The structural factor that was the incorporation of dog training into the nation's security apparatuses: it was in the best interests of the human population to carefully select and pro-actively manage the "biting" dog population, while increasing the number of "non-biting" dogs. One incidental factor was the growing unease with everything German in the years before World War II. The popular association of protection dog training with Germany was backed up with the origins of the "German shepherd" and the fact that many of the trainers active in the 1920s hailed from Most's police dog school.[8] Secondly, and more in tune with Howell's interpretation

8 The association of protection dog training with Europeans in general and Germans in particular (Josef Weber, Hans Tossutti, Carl Spitz, Willy Necker) was strong in the first

of "leash laws," increasing public concern with potential dog bites met with urban dog owners' growing desire to appease such fears, a goal that was easier to achieve through the rehashing of nineteenth-century ideals of canine loyalty and sagacity than through emphasis on dogs' aggressive potential. Displays of canine aggression had long been associated with the mostly illegal activity of dog-fighting; deemed abhorrent by respectable and morally-upstanding citizens and associated with members of the lower classes, recent immigrants (particularly Irish) and, as Bronwen Dickey phrases it in *Pit Bull: The Battle over an American Icon*, various "human riffraff" (Dickey 2016, 61). A heated debate on leash and muzzle laws took place in New York City toward the closing of 1929. The A.S.P.C.A. argued that muzzles were a cruel instrument of torture, while concerned citizens who sent in their opinions to *The New York Times* and *The New York Herald Tribune,* complained that the muzzle law was hardly ever observed anyway: "the average householder lets the dog out when he takes in the milk, and until night the animal leads a glad, free life" (Cleaver 1929, 18). It certainly did not help the case of the personal protection dog that many of the early police dog programs ended infamously when the trained dogs instigated fights with one another or bit their handlers (Chapman 1990, 16–19).[9] Finally, ownership of a personal protection dog signifies a lack of trust in the state's ability to protect the citizen, most certainly an element of the cowboy mythology of the "American Wild West" but one that seemed out of place among the urban New York bourgeoisie during the New Deal Era.

Two important voices in the discussion about how American pets should be taught to behave come from Most's two students: Josef Weber and Hans

---

half of the twentieth century and continued into the second half. The momentum for the formation of the United Schutzhund Clubs of America (1957) came from Czechoslovakian immigrant Gernot Riedel. The first trial was judged by Henry Friehs. The story of Willy Necker, whose books will not be discussed here in detail, is representative of how this association could actually help shape immigrant life stories. Necker was a working-class German immigrant trained as a carpenter who came to the US in 1929. He was unable to find work in his trade and took on a kennel job. His career took off and during World War II he trained dogs for the US Coast Guard. He is credited with importing the first Doberman into the US. Necker later maintained a kennel where he boarded clients' dogs and travelled with a troop of performing Dalmations. William Koehler, while American-born, was also of German descent.

9 Chapman cites a 1908 incident in South Orange, New Jersey: "On July 27, 1908, canine Bob was unfit for patrol because he was bitten by police dog Rover" (16) and recounts an incident told by Glen Ridge (NJ) historian Percy Suydan Wilson in which a policeman "failed to have the muzzle on the dog, and then accidentally stepped on the dog's foot. The policeman was taken to the hospital and it appears that he was fortunate to have escaped with his life" (19).

Tossutti. Weber, who was an influence on the founder of dog obedience Helen Whitehouse Walker and the trainer of Blanche Saunders, focuses on the inadequacies of the European model (as represented by Most but also, presumably by von Stephanitz and others). In his 1938 book *The Dog in Training,* Weber claims that "the European ideal (…) conflicted with the American desire for a loyal, well-mannered companion in a dog" (J. Weber 1938, 85). He explicitly equates this European ideal with the "foreign conception that a fully trained dog must have police and protective training" and emphatically repeats that these notions are incompatible with "American notions as to what makes a desirable and well-trained dog" (83). Weber even ends his book with a passionate plea against the training of protection dogs:

> However proud Europeans may be of the protective viciousness and power and training of their dogs, one thing is certain: We in America will not and ought not to change our fundamental attitude that, in private hands, such a dog cannot be tolerated. And I hope, too, for an early death of the absurd notion that every dog-owning citizen here thinks he can be his own police and train for his protection an animal that is a man killer. (199)

Hans Tossutti's book *The Companion Dog in Training*, published after World War II (in 1948) makes similar claims: "In this country we want the COMPANION dog, not the attacker. Protection trained dogs imported from Europe some years ago left in the public mind an unfavorable impression against one particular breed, and this aversion still has not been wholly overcome" (Tossutti 1948, 214, emphasis in the original).[10]

While both Weber and Tossutti repeat Most's assumption that dogs basically could not reason, they downplay Most's resulting emphasis on the manipulation of instinct as a training methodology. Their techniques, though not diametrically different from Most's, do not focus on developing that dog's protective and chase instincts. Instead, they see these instincts as in need of suppression. They also see the building up of instincts (particularly what Most

---

10 The use of the term companion dog should also be noted here. Nineteenth-century manuals routinely mentioned the training of pet dogs (or even "ladies' lap dogs"). In the 1930s the term companion dog gains a greater degree of prominence in connection with the popularity of obedience competitions, which result in the AKC-certified title of "Companion Dog." There is, of course, a clear symbolic dimension to the rise of the popularity of the term "companion dog"; as it connotes the dog's active and harmonious relationship with the human.

termed "chase" and "attack"), as responsible for the extreme violence of police dog training methods. Once a dog's instincts have been aroused, they have to be quelled through strong compulsion. In the eyes of these trainers, a way to avoid such extreme compulsion is not to develop these instincts in the first place. Weber writes: "in teaching a police dog his job, instruction is always accompanied by severe use of the whip. Deliberately we induce conditions that madden and annoy him to the end that in his rage to bite he will lose self-control and all sense of obedience. Then the whip, by arousing pain through its blows, must bring him back to obedience" (189). It is with Weber and Tossutti that the whip ceases to be seen as a staple of training equipment. However, it must be stressed that even for Most, the use of the whip is never a prescribed training accessory for eliciting a specific behavior (even the attack). It turns up only in what I have referred to as "the primal scene," that mythical confrontation that supposedly results in the permanent establishment of the human-canine hierarchy.[11] The whip is not used by Most in teaching the protection exercises because Most views the dog's submissiveness to the human as a prerequisite for defense work. Still, what is important for the argument that I am sketching is that Most's students did not blindly accept his ideas but were involved in the project of the "Americanization" of the methods important from Germany.

6 Conclusion

It is clear that the popular project of pet dog training becomes synonymous, at least in the minds of the trainers of the 1930s and 1940s, with a uniquely American project. This is not to say that the move away from teaching protection exercises to pet dogs does not happen elsewhere, but that in the United States it is seen as emerging in opposition to European influence. In the 1920s, aided with the new popularity of rugged masculinity and with the growing use of dogs by the uniformed professions, some Americans flirted with the notion of privately owned dogs serving the purpose of ensuring an individual human's security; with the dog beginning to serve the same function as a gun. While in this narrative the dog becomes a tool of hyper-individualized human self-sovereignty, the role of training in this endeavor remains similar to what it used

11 Most writes: "The switch should be employed until the animal submits and his will to resist, and the exasperation that accompanies it, is replaced by fear. So long as the dog does not submit, but continues to resist, a flexible switch should be used, if necessary, on the head and jaws, but not on the top of the nose. It may also be used on the neck, ears, legs, and tail" (26–27).

to be in the hunting manuals discussed in chapter 2. Training remains firmly embedded in the eugenic biopolitical framework and still serves as a tool of selective breeding. Most's establishment was, after all, also a huge state-operated breeding kennel. While training did not purport to measure intelligence, it did evaluate drives and their plasticity; specifically, how amenable the dog's drives were to human control. Dogs with developed yet plastic drives – those deemed strong and courageous but at the same time amenable to human control – were selected in breeding programs: not just in the German state-operated kennel but in American privately-owned kennels as well. In the 1920s, training accomplishments in "police dog trials" were routinely listed in American Kennel Club publications, dog show catalogs and training books as advertisements for specific dogs' stud services. The dogs who proved themselves in training were deemed worthy of reproduction and owning a trained dog was a way of increasing one's social standing. The training methods advocated by Most and his school of followers reflect that function of training: if training was to serve as a litmus test for how tough a dog was, the methods had to measure that toughness by eliciting confrontation. Anything else would miss the point. The violence, of course, also had a ritualistic element to it that corresponded with the then dominant vision of virile, white masculinity.

Most's project was a project in which training – fundamentally – could not become a mass undertaking: it still relied on the idea of elevating the subject of training from among the masses of untrained dogs. The trained dog was exceptionally tough and could become a dangerous tool in the hands of his human owner. In order for training to become an activity for "everydog," the practice needed to become part of a different project than the project of eugenic biopolitics: it needed to cease to be so strongly tied to breeding selection. This is the change advocated by Tossutti and Weber in their call for a truly American companion dog; the change most strongly associated with the figure of Blanche Saunders, though also, as chapter 5 suggests, with the ideas of B. F. Skinner.

CHAPTER 4

The Rise and Fall of Obedience: From Helen Whitehouse Walker to the Dawn of Positive Training, 1933–1984

> So we should not see things as the replacement of a society of sovereignty by a society of discipline, and then of a society of discipline by a society, say, of government. In fact, we have a triangle: sovereignty, discipline, and governmental management, which has population as its main target and apparatuses of security as its essential mechanism
>
> MICHEL FOUCAULT, *SECURITY, TERRITORY, POPULATION*

•••

> The object of police is almost infinite. (…) [T]here is no limit to the objectives of government when it is a question of managing a public power that has to regulate the behavior of subjects.
>
> MICHEL FOUCAULT, *BIRTH OF BIOPOLITICS*

It seems almost a cliché to speak of dog obedience training as Foucauldian disciplinary power applied to non-humans. All the features of the disciplinary regime are apparent in the practice as it is remembered from the second half of the twentieth century: it is tied to the rise of institutions (the dog training school); it is directed at the body, molding it into docility through detailed procedures (for pressing, pushing, jerking); it is necessarily corrective (choke collars and leashes); and its goal is increasing the body's utility and docility. It relies on surveillance, normalizing judgment and the examination: the leash, the crate and the obedience trial. It is tied to the presence of the body in a series of distinct spaces: the home, the school, the obedience ring. It manages a vast multiplicity by relying on individualizing. In the case of pet dog training in the US, the refinement of disciplinary techniques associated with the rise in popularity of obedience training is accompanied by a major shift in the function

 | DOI:10.1163/9789004380295_006

training played: rather than remaining an elitist tool for proving the superior intelligence of an individual dog, or a tool for breeding selection, dog training becomes a popular practice embedded in the project of governmentality, that is of ensuring the safety, health and overall well-being of the urban human population. While this project was highly anthropocentric, its side effects – somewhat unintentionally – paved the way for reinforcement-based training and the ensuing popularity of Skinner's behaviorism among dog trainers at the end of the twentieth century.

The context of the story of the rise of the disciplinary regime in relation to these specific non-human animals in the US is also significant: obedience training is a practice that develops at a specific time (1930s and 1940s) and place (urban and semi-urban Northeastern US). Most interestingly, this development of disciplinary practices targeted at dogs is tied to shifting associations with (human) gender. While Foucault's docile body is always a male body (the running example in *Discipline and Punish* is the body of the soldier), most famous feminist applications of Foucault to women focus on women as potential subjects of discipline. Women who function as builders and agents of the disciplinary regime have been given less thought. In this story, it is two women who are largely (though not single-handedly) responsible for the strengthening of the disciplinary character of dog training. Helen Whitehouse Walker and Blanche Saunders are representatives of the kind of dog training advocated by Weber and Tossutti as befitting the American nation: the training of the American companion dog; the dog who does not bite, does not pull on lead, stays, comes, sits, does not chase cars or squirrels, does not jump up on people and does not bark when left alone. Yet it would be a gross oversight to speak of Walker and Saunders as just dog trainers, without regard to their gender and social status. The fact that they were white women – one of whom was also quite wealthy and well connected in New York society – is absolutely fundamental for the role they played in the development of dog training, though the influence of their gender on the activity they were involved in was not Walker's intention; neither was the scale of the popularity of training and the function it began to serve. In fact, a huge discrepancy can be identified between Walker's goals and the actual developments in the field. Walker, an upper-class socialite from Mount Kisco, NY, wanted to use the rhetoric of usefulness associated with dog training in the early twentieth century to rescue her beloved poodles from the label of "spoiled lapdogs." Obedience trials were supposed to serve as a tool for the breeding selection of urban dogs: an alternative to the show ring and to the field trial. Walker never achieved this goal; in the public eye poodles have remained fancy lapdogs but through the explicit and highly publicized involvement of women in the sport, dog training lost its masculine aura and gradually – this did not happen overnight – began to be associated with qualities deemed feminine: care, sensitivity and love. Walker was only involved in the setting up of AKC obedience in the 1930s; she later handed over the field – and even her own kennel name – to Blanche Saunders, who remained a popular (possibly, the most popular) American trainer until her untimely death in 1964.

The switch from upper-class Walker, whose goals were still quite elitist (to prove the superior breeding of her poodles), to working-class Saunders, who finally taught the first obedience class open to mixed breed dogs has an obvious symbolic dimension. This is the period when training becomes a democratic endeavor, involved in the project of fostering beneficial human-canine relations. On the declarative level: it ceases to be just about making the human's life easier but begins to be viewed as necessary for making the dog's life among humans easier. For Most, training benefited the human and entailed a necessary sacrifice on the part of the dog wo gave up his freedom to "serve" mankind and the nation. Not so for Saunders. In *Training you to Train Your Dog* (1965, first edition: 1946) Saunders writes: "Training is one way of caring for his [the dog's] welfare. It teaches him manners; it will help to keep him in good physical condition, to have steady nerves and to become a more congenial and a more amusing pet and companion" (Saunders 1965, 12). While this approach may seem similar to that expressed by nineteenth-century pet trainers, there is a fundamental difference. The Victorians, both in England and the US bourgeoisie, argued that training is beneficial for the dog because it makes him almost human; by elevating him among others of his kind, offering a promise of "transcending animality." In the nineteenth century trainability was still tied to intelligence or sagacity; a link that was later destroyed by Konrad Most. This approach implied the existence of those others who are not elevated as a result of training, who remain "dumb creatures" and against whom the exceptionality of the subject of training can be measured. The kind of project represented by Saunders sees the benefits of training in regulating human-animal cohabitation, not in a possibility of the transcendence of animality. It is not an expression of human benevolence; it is necessary for the human's safety and comfort of living in a crowded world populated by members of different species. Even more significantly, if the project is to work, it must be addressed to "everydog," not just to the selected one.

Dog training – as envisioned by trainers like Saunders, Weber and Tossutti – begins to serve as a tool of governmentality; an element of the management of the ever-growing urban and suburban canine pet population, seen as impacting the non-human population's well-being, health, but primarily as mitigating the pet population's potential negative effects on the human population. In order to serve that purpose training must become widespread and easily available to pet owners. As a result of attending classes, dog owners should begin to internalize the rules for responsible dog ownership and to self-regulate the behavior of the dog-human pair meandering through urban space. Saunders goes on to argue that: "When training has become an accepted part of every dog's education and when owners, as well, realize their responsibility, there will be less need of city and state legislation to control our pets" (Saunders 1965, 12). In connection to such views, this period is also one of intense codification, standardization and institutionalization of dog training, accompanied by its growing popularity among urban and, later also suburban, dog owners. This is the era of the emergence of the dog obedience school, which operates like a classic Foucauldian disciplinary institution: instructing, correcting, evaluating progress. This is the era of formalization and normalization: dog training becomes an activity practiced through an institutionalized framework, whether these institutions be dog training clubs serving the pet owner or publicly and privately-operated institutions for training military dogs or seeing eye dogs. The effectiveness of training begins to be judged through standardized examinations, which range from graduation ceremonies for obedience class attendees to formal obedience degrees or state-certified examinations. This is the period when obedience trials are first held in the United States (1933) and become sanctioned by the American Kennel Club (1936).

The obedience trained dog begins to be produced on a mass scale through methods that are developed specifically with this purpose in mind: they are meant to be applied easily to a large number of dogs. William Koehler's book, *The Koehler Method of Dog Training* (1976, first edition: 1962), begins with an affidavit certifying that in the period 1946–1962 Koehler trained 9,500 dogs as the director of the obedience program of the Orange Empire Dog Club (Koehler 1976, 7) and the back cover boasts that Koehler personally supervised the training of 15,000 dogs. Saunders also bragged about the sheer number of dogs that went through her hands, as did other trainers to establish their credentials. Additionally, Saunders authored the first textbooks for obedience school instructors, taught seminars to potential obedience instructors and promoted the idea of dog obedience clubs; where more experienced members teach classes to others. In *Dog Obedience: The Guide for Trainers* (1954) Saunders argues that that the size of the class is limited only by the size of the facilities available for instruction. As the popularity of training grows, facilities become more easily available because various authorities recognize the social significance of the project. Saunders notices that the general public is beginning to recognize "the benefits derived from such activity" and lists instances when "authorities have been persuaded to open the doors of school gymnasiums, community centers and city halls" as wells as city parks, YMCAs, university campuses, riding stables and even restaurants to training classes (Saunders 1954, 20). Dog training is seen as a civic duty that all should become involved in.

Training becomes the mechanism for the production of docile bodies and docility is defined quite literally here: bodies that will not bite; bodies that will not maim human handlers and that can co-exist with human beings and other animals in an urban context without posing a threat to their surroundings. This is quite different from how training was understood in the opening of the twentieth century and within the "masculine" tradition that can be traced past the 1920s in mainstream pet training via William Koehler and his followers. Despite not sharing Most's ideas on training through developing instinct, Walker, Saunders, Weber and Tossutti all share Most's views on canine intelligence: they believe it is overrated. All their textbooks are also steadfast in disassociating intelligence and trainability. Saunders does not see intelligence as in any way related to the effects of training: " 'Can every dog be trained?' Every dog will improve with training, but, naturally, some learn more quickly than others. The breed of the dog, the age of the dog, his disposition, and the owner's ability as a trainer are factors that determine how successful training will be" (Saunders 1965, 9). Meanwhile, Tossutti recounts the mistaken – in his eyes – views of the past, only to correct them: "I do not believe there is any such thing as reasoning power in the dog. I believe, rather, that he merely reacts to directions given by man" (Tossutti 1948, 19). Weber concurs, adding that it is in fact the dog's lack of reasoning skills that makes training essential: "Just because he is only a dog and has no highly developed reasoning power but possesses only instinct, we, to achieve a better mutual relationship, must remain man and seek to appreciate and understand the dog from the standpoint of his own lower level of development" (J. Weber 1938, 15). This curious about turn in the relation of training to intelligence, evident already in Most's manual but present in American pet dog training manuals through the 1980s, ceases to be surprising when one reads it in relation to Foucault's understanding of discipline: disciplinary techniques are meant to be efficient; useful on a wide scale and their success does not depend on the intelligence of the subject of discipline.

1 Leading Others: Tools of Discipline

Philip Howell's ideas about the shift from the technologies of discipline to those of governmentality in managing the presence of dogs in public space in Victorian England (2015) can be helpful in understanding the democratization of training in the US since the 1940s, as well as the decline of acceptance for public displays of canine aggression. If training was to help in peaceful human-canine coexistence, it could not develop a dog's aggressive potential and – again – it had to become popular. The view of training as a tool for governing entire populations of dogs – largely in order to safeguard the human population from dog bites, maulings and to increase the overall hygiene in cities – becomes widespread, as evident from this newspaper snippet about the first obedience class put on by the A.S.P.C.A in 1944 and taught by Saunders:

> It goes without saying that New York City's dogs need training. The recent rabies scare in the Bronx would be less fearsome if there were more dogs on leashes and fewer running around loose in direct violation of city laws. Hundreds of accidents, causing injury or death to city-owned dogs, could be avoided if dogs could be taught to walk at heel, to return to command and to be obedient in leash. (Patterson 1944, B2)

What transpires from this story is the common conviction that training is not so much about teaching behaviors to dogs as it is about teaching responsible dog ownership to humans, as Howell wrote, using Foucault's terminology, "the conduct of conduct," which consists of both " 'lead'[ing] others (according to mechanisms of coercion which are, to varying degrees, strict) and a way of behaving within a more or less open field of possibilities" (Foucault 1982, 789). The primary understanding of the goals of training in the above excerpt is to bring dogs and humans together: very literally, through the use of the leash or lead, which – coincidentally – also happens to be the word used in the English translation of Foucault's essay "The Subject and Power": "leading others." However, the leash, which for Howell is the symbol of the shift towards governmentality, is also very much a tool of discipline when it comes to actual training technologies; when we analyze it not as a symbolic representation of the sphere of "the conduct of conduct" but as a practical tool for shaping a dog's behavior.

The connection with discipline as a method of governmentality does not contradict Foucault's understanding of the relationship between the two different modalities of power, as the passage from *Security, Territory, Population* which serves as the epigraph for this chapter proves. The public perception of training as a tool of governmentality does not, in any way, preclude the use of disciplinary techniques in actual human-canine training interactions. In fact, the leash is a much better representation of discipline than the muzzle, which does not alter the dog's behavior but simply provides a physical barrier between the dog's potential display of animal aggression and the human skin. While organization of space in a way that isolates the delinquent is a disciplinary technique, altering behavior to create docility is just as much a trademark of the disciplinary regime. The muzzle is not directly a tool of training (even though it could be seen as part of the training apparatus, to recall Michael Peterson's term), while the leash was the most significant tool in the trainer's toolbox in the type of training under discussion here. Training off-leash would have been inconceivable within a methodology based on physically influencing the position of the dog's body. In fact, the way that the leash is used by Saunders connects it to Foucault's classic descriptions of the features of disciplinary power from *Discipline and Punish*. The leash is both a tool of observation and surveillance but also a corrective measure of normalizing judgment.

All formal training takes place on lead, which is necessary for executing the training techniques proposed by Saunders and other trainers who shared her methods. Even the shift to the contemporary "heel" position, as described in the previous chapter, was motivated with the need to supervise the dog's presence more closely. The leash is also used to physically prevent undesirable behaviors (chasing cars, squirrels, etc.) and to confine the dog when the owner cannot supervise him directly. In formal training, the dog's presence in the correct "heel" position is secured through the use of the lead, which also serves as the tool for delivering corrections should the dog leave the desired position. The transition to off-lead work is a gradual one, consisting of a number of steps in which the handler deceives the dog into believing the human still has the possibility of correcting the dog, even if the dog is not on leash. As a first step, the leash is placed on the handler's shoulder (to discourage the trainer from overhandling it, while still providing control of the dog); it is then exchanged for a light clothesline that tricks the dog into thinking he is free and yet allows the handler to execute corrections, only to be eliminated once the dog has become so proficient at performing the exercise that corrections are no longer needed (Saunders 1965, 166–168). If we try to theorize how this effect of self-regulation is achieved, then it is most certainly through almost classic Panopticism: the dog comes to believe that his movements are still controlled by the leash, even if they are not. Saunders's advice for reacting should the dog become aware that the handler cannot correct him is simple: put the dog back on leash and repeat the exercise. Disciplinary techniques target the body and the dog begins to behave "correctly" out of habit.

Saunders, similarly to Most, reminds her readers to maintain a significant degree of slack on the lead to allow the dog to make mistakes (102), which are then corrected with strong jerks on the leash. Interestingly, the writer whose ideas about the human-canine bond seem the most progressive, Hans Tossutti, makes training less physically strenuous for the dog by keeping less slack on the leash. The dog never has the possibility of straying too far from the owner and the corrections, per laws of physics, are not as strong. On the other end of the spectrum, William Koehler – rather infamous among contemporary trainers and writing in the tradition of Konrad Most – begins his training program of teaching the dog "attentiveness" by using a long line. If the dog strays too far from the owner, the trainer is advised to execute an about turn and move away from the dog. Clearly, the power of the correction thus delivered is much stronger than the one administered by Saunders or Tossutti. Koehler's reasoning is that a stronger correction will eradicate the undesirable behavior quickly and efficiently, without the need for multiple repetitions of corrections and without – what he calls – "nagging." Still, the measure of gentleness within the dominant teaching style of the period can only consist of establishing the severity of the physical correction: with Tossutti on one end of the scale and Koehler on the other. Saunders's methods are somewhere in the middle but are not fundamentally different from those of Koehler. She still relies on inducing mistakes and then physically correcting the dog, increasing the strength of the correction relative to the (perceived) seriousness of the dog's misbehavior. With a dog who protests on-lead heelwork by lunging at the hander, Saunders suggests a radical approach:

> The hysterical type of dog will growl and jump around and climb all over the trainer when he finds himself restrained by the leash the first time he is made to heel. The dog may even lunge at the trainer and try to bite. A sharp cuff on the nose, which is the usual correction, will only make him worse than ever. Instead, he will have to be handled severely and firmly in another way. He should be lifted up momentarily by the leash and held at arm's length with his front feet in the air. When his anger subsides, he should be lowered quietly to the ground and then patted. (Saunders 1965, 104)

There is absolutely no feminine softness in Saunders's solution to the dog's reluctance to follow the trainer; not to mention the fact that she does not completely forego the whip, an innovation that is sometimes erroneously attributed to Saunders, though she does not use it for eliciting behaviors, but for stopping bad habits deemed life-threatening to the dog and those around him; for example, for chasing cars, cats and children (61). In similar situations, she also resorts to throw chains and BB guns (63). While Saunders advocates against kicking a dog in anger, she routinely uses a slight kick to the dog's rear end to speed up the about turn and a kick to the dog's chest to teach the left turn in heeling: "If at any time the dog walks behind the trainer, or if he cuts in on the trainer's right side, a quick but careful kick backward with the right foot or a cuff on the nose with the right hand is the correction" (106). In fact, in an illustrated article for *Life* (1940), Saunders is quoted as saying: "He who kicks a dog kicks his own soul toward hell" (*Life* 1940, 69) directly above a picture where she is shown giving a "kick correction." This distinction between physical violence, which is to be avoided, and "corrections," which are necessary, is an echo of nineteenth century ideas about the value of self-restraint.

The distinction between how Saunders and Koehler are remembered by contemporary dog trainers speaks volumes about the relationship between gender and the specular logic of violence. William Koehler, one of the most famous trainers of the 1960s, is today associated with the negative legacy of correction-based training. Saunders, on the other hand, is much less controversial, even though, as previously noted, there is no strong qualitative difference in the methods employed by the two trainers: they both relied on correcting misbehavior, using the leash and the choke collar as primary teaching tools. Still, there is a sharp difference in how Saunders writes about administering corrections. While Koehler is also explicitly involved in the disciplinary project of creating docile canine bodies, there is a certain excess to his ideas, his vocabulary, his rhetoric. Even though Saunders's methods can sometimes come very close in their physical brutality to those of Koehler, their violence is subsumed within the bigger narrative of working towards peaceful co-existence. For example, in dealing with a stubborn dog who does not want to submit, both trainers advocate that the dog should be choked by being lifted into the air. Saunders explains the need for lifting the dog off his feet and choking him with a choke collar very matter-of-factly in about two lines: lift, choke, put the dog back on the ground, praise and carry on (Saunders 1965, 104). The violence of the method is subsumed within the efficiency of all of the trainer's actions. For Koehler, his signature "helicopter" maneuver seems something to be relished, drawn out and described in vivid detail, similarly to the use of "the tranquilizer": a rubber-covered wooden dowel used to strike the dog on the muzzle (Koehler 1976, 54). The description of how to prepare for using the dowel reads like a description of preparing for battle: if one suspects the dog might bite the handler, "the operation has to be carefully planned;" the dowel is placed in just the right position: hidden in the handler's pocket, under his belt or behind the handler's ear. The handler works the dog at heeling "until the dog shows the first sign of resentment" (53) at which point the "major operation" begins. It is this celebration of violence – this apparent joy of going to war with the dog – that appalls Koehler's opponents today. Violence is used, as it seems, not just with the goal of increasing the body's productivity and usefulness but is highly ritualistic; a reconstruction of the "primal scene" of man's conquest of nature, that seemed fitting in the 1910s but that stood out against the current of modernization in the 1960s.

2 Governmentality

I have repeatedly mentioned the shift of the function of training as it becomes embedded in the framework of governmentality. Yet, additional clarification may be necessary because governmentality and biopolitics sometimes function as synonyms in discussions of Foucault's work. While neither concept is ever explained in a fully cohesive manner, governmentality seems to replace biopolitics as the preferred conceptual framework towards the end of Foucault's life. The most explicit definition of governmentality is provided in *Security, Territory, Population* (2007) though the interviews from *Power/Knowledge* (1980) also shed light on Foucault's understanding of the concept. In both texts governmentality is seen as a power that affects the entire population rather than only the body of the individual subject. In *Security, Territory, Population,* Foucault speaks of governmentality as a "power which has the population as its target, political economy as its major form of knowledge, and apparatuses of security as its essential technical instrument" (Foucault 2007, 107–8). While the resemblances between biopolitics, which is also aimed at the population, and governmentality are striking, I choose to speak of a shift from biopolitics to governmentality because maintaining this differentiation helps conceptualize the change in the function that the activity of dog training plays: it goes from being involved in a biopolitics of the canine population to being a strategy of the governmentality of the human population. Along the way – slowly and somewhat unintentionally – dogs also begin to be addressed as potential subjects of governmentality themselves. We go from training being a selective measure for shaping the population of dogs to training as safeguarding the human population; safeguarding them from dog bites, from being knocked down by rambunctious canine greetings, from having the sofa chewed up, etc. As this happens, the focus on training as a tool of breeding selection (particularly for the pet dog population) decreases.

Some Foucauldian scholars prefer to retain the term biopolitics to speak of interventions in the biological make-up of society, while using governmentality to address other strategies of affecting the well-being and make-up of the population. Nikolas Rose speaks of "eugenic biopolitics" (2001, 2009) to refer to the early twentieth century strategies of affecting the biological make-up of the population that focused on encouraging or discouraging reproduction among various groups of the human population. This term seems best suited to addressing the strategies applied since the mid-nineteenth century to the management of "purebred animals." In my reading of the changing role of dog training, the practice was first embedded in a eugenic biopolitics of the canine population: it served as a selective measure in the process of breeding. Dogs that were deemed the most trainable were reproduced. I have discussed how this type of eugenic biopolitics saw itself as rooted in the science of biology, even though it relied on culturally constructed categories. However, as the twentieth century progresses, dog training becomes an obvious part of the various strategies of addressing the well-being, health and hygiene of the human population. This shift is also seen, as is obvious in the writings of Saunders, as benefitting the canine population but this seems to be a side effect of the undoubtedly anthropocentric endeavor: in a world organized around the needs of humans, it is safer for dogs to curb their animal instincts. In *Power/Knowledge* Foucault speaks of the "disposition of society as a milieu of physical well-being, health and optimum longevity" (Foucault 1980, 170). If we accept the premise that humans and dogs are going to live together in close quarters, this coexistence might as well be made easier and safer. Foucault sees control of urban space as a key element of governmentality because it constitutes "perhaps the most dangerous environment for the population" (190) and it comes as no surprise that dog training is promoted primarily among the urban population. City dogs (though also increasingly suburban dogs in the post-World War II reality) are the dogs that interact with the largest number of humans on a daily basis, and it is their behavior that needs to be controlled most closely. All of this is not to say that dog breeding suddenly exited the framework of eugenic biopolitics. It did not and, in fact, still has not to this day: humans obviously still exert control over canine reproduction but, at least in the case of pet dogs, training accomplishments are hardly ever a criterion that is taken into account. Trainability remains significant only outside of mainstream breeding practices aimed for the pet market: that is among trainers/breeders of sled dogs, herding dogs, hunting dogs, etc.

This shift was not envisioned by a single individual: in fact, it went very much against Helen Whitehouse Walker's vision of dog obedience. Blanche Saunders's historical look at the emergence of competitive obedience trials opens with a 1937 article from *The New Yorker*, titled "A Trend in Dogs." The article is a satirical representation of a conversation between two female dog breeders who comment on the new trend of using obedience trials as a criterion for breeding selection. The two women oppose the burden that the new trend could potentially place on breeders. One of them complains: "As soon as this insane thing is upon us, my only hope will be to start breeding for intelligence. [. ...] It's going to be frightfully uphill work" (Saunders 1974, 25). In addition to revealing that *The New Yorker* considered dog training a social phenomenon significant enough to warrant satire, the article accurately reflects Helen Whitehouse Walker's original intentions for the sport of dog obedience. Though Walker would not have used to word intelligence – she was too directly an intellectual descendant of Most (via Josef Weber) – she did see the new activity as part of the old framework of eugenic biopolitics: certainly, as an improvement on using only physical appearance as the sole criterion for breeding selection of purebred dogs. The incorporation of the military-style exercises propagated by Most and the steadfast rejection of exercises even remotely resembling those routinely taught to pet dogs in the nineteenth century are also clear indicators of Walker's program: she wanted to build the prestige of dog training as a hobby by associating it with the rhetoric of utility and disassociating it from entertainment (via the circus legacy).

This insistence on the usefulness of heeling and the uselessness of rolling over is almost obsessive and becomes written into the first AKC regulations for obedience trials: "The idea of Obedience Tests is to demonstrate the usefulness of the pure bred dog as the companion and guardian of man and not the ability of the dog to acquire facility in the performance of mere tricks" (Saunders 1974, 93). Foucault's words on the necessary coupling of usefulness and docility from *History of Sexuality, Volume I* provide a valid explanation for the obsession with utility (there is no point in docility outside of a framework of utility), but it is clear that the emphasis Walker constantly placed on utility was actually a reflection of the anxiety she must have felt about the arbitrariness of the exercises chosen and the purpose they were supposed to serve. Advanced obedience exercises – in addition to simple heeling, coming and staying – included an urbanized version of field trial tasks: retrieving, sending a dog to a desired location, scent discrimination and jumping over a hurdle on cue. To a contemporary ethologist it is obvious that a dog's success in retrieving a dumbbell is not an indication of his field performance and most hunters would laugh at the idea of using the results of obedience tests as tools in a breeding program of actual hunting dogs. In fact, the general public perceived Walker's dogs as "trick dogs," as evident in the scores of newspaper and magazine articles devoted to the topic in the 1930s and 1940s. In other words, the public saw through the smokescreen of utility, but still responded positively to the idea of obedience competitions and obedience training. I hypothesize that this positive reaction resulted from the perception of the activity as involved in the new project of governmentality, not in the project of eugenic biopolitics that Walker was openly advocating.

At the same time, while Walker most certainly was interested in spreading the popularity of the sport of obedience and used her social connections to help ensure interest, she most likely did not envision the activity as expanding beyond the environment she knew as the dog circles: the upper-middle class breeders of purebred dogs of non-working breeds, largely (though not exclusively) female. She most certainly made huge efforts to increase the visibility of training, but these efforts reveal her to be working within a set of rigid constraints and show how blind she was to the situatedness of her own efforts. For example, in 1937 Walker and Saunders set off on a coast-to-coast "trailer trek" from New York to California and back, with the goal of popularizing obedience competitions by giving obedience demonstrations at AKC shows across the nation. While it is difficult not to admire the sheer courage of two women travelling cross-country without male escort, in an automobile, while hauling a trailer with three dogs, it is also difficult to not gasp at the willful ignorance of the women regarding the economic circumstances of the communities they were travelling through. Blanche Saunders kept a travel log during her journey with Walker and summarized it in a chapter of *The Story of Dog Obedience*. It is striking that the world outside of dog shows is completely missing from the account, as if the entire endeavor was designed in a way that allowed the two women access only to an alternative America: prosperous, content and preoccupied solely with the qualities of purebred dogs. The Dust Bowl only figures in the chapter as annoying weather that presents the women with the choice of "choking to death with the windows open or suffocating with the windows shut" (Saunders 1974, 111). Saunders and Walker also brag about how the poodles "became experts at finding hobos asleep under a bush and would chase them with fierce determination" (113). Walker's project is clearly not one of social change.

After devoting several years of her life to propagating obedience trials and obedience training (of purebred dogs), Walker abandons the endeavor in a decision that she never explains directly, and Saunders takes over as the major force of the movement. Saunders seems much more suited to the task of doing the legwork actually required to promote the activity, and she works tirelessly for several decades to train dogs and trainers, explicitly professing goals of governmentality, not those of eugenic biopolitics. She opens *Training You to Train Your Dog* with the claim that training has become "an accepted part of every dog's upbringing, essential to his well being" (Saunders 1965, 1) and adds: "purebred or not, every dog deserves to be trained just as every child has the right to an education" (3). Saunders runs classes for the A.S.P.C.A.; she supports the 4-H dog training program and in the last photograph that was taken of her she is shown as teaching a group of girl scouts.[1] Of course, the question remains what happens in the fifth decade of the twentieth century that makes the idea of incorporating dogs into the framework of governmentality appealing? Why does the idea of training ALL dogs to walk on leash and sit politely catch on at this particular time?

The answer is obvious, though it may at first seem counterintuitive: war and the social changes that followed after the war, especially suburbanization and urban sprawl. World War II marks the first time that the US used dogs in war in an organized way. More importantly, it marks the first time that a coordinated attempt at training large numbers of dogs was carried out, via the Dogs for Defense project. Forty thousand pet dogs were donated and 8,000 of those successfully completed training and were deployed (Alger and Alger 2013, 84). Saunders actually participated in the training of these military dogs, but it is not her personal involvement that matters. More importantly, the success of this massive training effort for military purposes paved the way for training pet dogs on a mass scale and in an organized way. On the meta-level, Dogs for Defense proved a successful incorporation of dogs into the security apparatuses of the state. Canines became a tool of the police, in the understanding of police that Foucault proposes in *The Birth of Biopower*, that is as an institution of social regulation; not of justice, but of normalization. It is not that dogs had not been beneficial to humans before, it is that they become incorporated into the apparatuses of governmentality: the behavior of dogs begins to be seen as important for the functioning of the nation in a way in which it had not been before. It is not coincidental that the military effort was accompanied by the increasingly common use of dogs by the police and by the emergence of the guide dog program. Once the state understands that shaping the behavior of canines can be an element of military policy, it is not much of a mental leap to make the connection that shaping the behavior of pet dogs can be used within the same framework of governmentality. It is in the interest of the state that military dogs detect mines, police dogs attack on cue, and pet dogs walk nicely on leash and allow themselves to be petted. The disciplinary methods and institutions associated with Most are transformed in a way that makes it possible to achieve that goal. America's successful deployment of dogs in

1 The photograph is on page 189 of *The Story of Dog Obedience*. The story of 4-H's involvement in dog training is fascinating in itself. 4-H begins to offer dog training programs in the late 1940s to young people living in the newly formed suburbs; that is, to ex-urbanites. It is in the post WWII decades that dog training is added to animal husbandry in the 4-H programs, at the same time creating a new audience (suburban youth) for the organization, which traditionally had been traditionally active in rural areas of the US.

World War II – and specifically the training program developed for military purposes – can begin to serve as a template for the training of pet dogs. The fact that many trainers who were involved in training dogs for the military during World War II (Koehler, Saunders, Weber, Spitz) later become the faces of the obedience training movement further strengthens this claim.

3 Training You to Train Your Dog: Layers of Human-Canine Discipline

I have suggested that discipline is the primary tool of putting this project of governmentality into practice: the way of ensuring the safety and well-being of the human population is through the exercise of disciplinary techniques on the bodies of individual canines in a group context. The techniques advocated by Saunders share almost all the features of the disciplinary techniques described by Foucault in *Discipline and Punish* and the disciplinary institution she advocates, the dog training school with its group classes and regular lessons, is almost a mirror reflection of the classic disciplinary institutions for humans. However, there is a certain specificity in the way that discipline is exercised on canine bodies that is, on the one hand, a result of the involvement of training in governmentality and, on the other, an interesting premonition of the changes to come. The title of Saunders's first book, *Training You to Train Your Dog*, suggests that the complex of power/knowledge is more multi-layered than in the case of disciplinary institutions for humans. While discipline is impersonal, it is practiced by experts directly on the bodies of human subjects. In the case of dog training classes, the instructor teaches the human trainer who, in turn, trains the dog. This practice of discipline in the institution of the dog training school does not target simply the body of the dog. There exists an intermediary between the expert and the animal: the dog's owner.

The significance of involving the owner of the dog in the process of training cannot be overemphasized. In the 1940s it is everywhere: in the titles of instructional books and newspaper articles; in the title of the first instructional video on dog training: Blanche Saunders actually wrote the script for the movie *Training You to Train Your Dog*, narrated by actress Helen Hayes and broadcaster Lowell Thomas.[2] The A.S.P.C.A.'s promotional materials distributed before the start of the first dog training class open to all dogs, regardless of ancestry, advocated that: "Dogs and owners are always trained together, the theory being

2 At the time of writing, January 2017, the movie is publicly accessible on Youtube, on the iCandyTV channel. https://www.youtube.com/watch?v=mxHTfdi5jLw&t=1220s October 27 2015, accessed January 24 2017.

that we teach you to train your dog" (Lane and Zawistowski 2008, 91). Similar views were expressed in articles published at the time in the popular press. In the story of dog training the idea that it should be the owner who trains the dog is promoted as an innovation of the 1940s, in opposition to the supposedly popular practice of sending dogs off to professional trainers. As Lane and Zawistowski write in the history of the A.S.P.C.A.: "What was unique about the new obedience was that the dog's owner was taught how to train and compete with her own dog" (92). While it is not true that the training of pet dogs by their owners comes around only in the 1940s – chapter 1 makes it clear that training the household pet to perform tricks was seen as a proper educational undertaking for young people already in the 1860s – it most certainly is true that organized classes become popular only after World War II. It is also telling that in the story of dog training the owner's involvement in the process of the dog's education is seen as a modern development: it corresponds neatly with the shift to governmentality. The idea of sending pet dogs off for training to professional trainers was actually only popular in the US among the upper echelons of society, maybe with the exception of the massive Dogs for Defense effort but in that case dogs were trained for the military.[3] Thus, what changes is not really the idea that training should be carried out by the dog's owner but the scale of the training enterprise and the environment where training happens: a group class taught by an expert.

In Philip Howell's analysis of leash laws as a tool of governmentality in Victorian England, these regulations are discussed as aimed at the human-canine dyad; as involved in a strategy of regulating the behavior of the dog through the behavior of the owner. This goal is even more visible in the work of dog training classes: while the dog is the recipient of disciplinary techniques administered by the owner, the owner's position as intermediary does not leave him or her unchanged. The obvious purpose for actively involving owners was to imbue in them a sense of responsibility for the behavior of their own dogs. In her textbooks for dog obedience instructors, Saunders writes: "The owner who enrolls his dog in a training class assumes definite responsibilities. He has a duty to the club, to the training director, and to his dog" (Saunders 1954, 35). Training classes were not taken lightly at the time: Saunders discusses severe penalties – such as withholding graduation certificates – for missed classes

3 Having a dog trained by a professional trainer was something that many sportsmen resorted to in the nineteenth century. Similarly, some shepherds tended to offer "started dogs": dogs with basic herding training. According to advertisements in the breed magazines of the 1920s and 1930s professional trainers worked with "working dogs" (German shepherds, Dobermans), which – assumedly – were taught protection work.

and procedures to follow should a female dog come into season (the owner should still attend and observe the classes). It is interesting that the duties Saunders discusses were not simply duties towards the dog: these were primarily duties towards other humans. In addition to developing the feeling of responsibility for the behavior of one's dog, turning dog training into a social endeavor created a self-policing community bound by bonds of camaraderie and shared experience.

On the other hand, it seems that a side effect of the close human-canine interaction in a structured and formal environment was a redefinition of the bond between the human and the canine, perhaps because both the human and the dog were quite explicitly subjects of power in the arrangement of this disciplinary institution. After all, in the title of Saunders's book, the human is addressed using the same vocabulary as that applied to the dog: *Training You to Train Your Dog*. Many of the experts insisted that training the human was more difficult than training the dog, because the humans came to class with the wrong mindset: one which had to be changed in order for training to take place. William Koehler writes of the prevalent misconception that dogs "want to please" and many other instructors wrote of the need to work through the dog owner's anthropomorphic misconceptions. In a 1976 article for the *Chicago Tribune* Rhona Schultz thus sums up multiple conversations with prominent Illinois dog trainers of the 1970s: "Most trainers agreed that the dog owner is far more difficult to train than the dog" (Schultz 1976, F1). The difficulties in training the owners included their persistent leniency in administering corrections and overall disdain for following procedure.

Already in 1962 when Koehler published his book, which actually won him an award from the American Humane Association, he derided owners who were "wincers" (Koehler 1976, 14), that is those who found the methods too cruel and needed to be trained themselves. In a style unusual for a training manual, Koehler seems somewhat angered by the owners who disagree with him and sets his sights on retraining them. Writing in third person, even though the rest of the book uses the personal voice, he describes how the instructor evaluates the owners as they arrive for the first class with their dogs and singles out those in need of training: "The instructor had been studying and classifying the people and their dogs, working to prepare a format he would use to prove his point – that intelligent dogs rarely want to please people whom they do not respect" (18). In fact, the entirety of chapter 1 of Koehler's book is devoted to replying to criticism of his methods. It seems a likely explanation that the discomfort many dog owners began feeling when applying the training methods prescribed to solve their problems resulted from empathizing with their dogs. Somewhat ironically, however, it was the turn to governmentality – with its emphasis on the human-canine dyad, concurrent strengthening of the human-canine bond and the dissociation of training and breeding – that initiated the discomfort with classic disciplinary techniques.

Of course, it would be an overstatement to claim that the very coming into being of training classes redefined the human-canine relationship in a way that made using the word bond appropriate when discussing the subject. Literature documenting the changes in the human-canine bond since the 1980s into the present is overwhelming, but rather unanimous in its acknowledgment of the growing significance of the emotional attachment between the individual human and the pet dog: the gradual creeping in of the pet into the contemporary apparatus for kin-making. Among myriad other factors, the popularity of dog training most certainly contributed to this redefinition of the human-canine bond: taking one's dog to a class, usually by car, certainly added to the amount of time the human and the dog spent together and to the diversity of experiences they shared, including the experience of being subjected to disciplinary techniques that adults only vaguely remembered from the times of their childhood. It seems as if the disciplinary regime as applied to dogs carried within itself the seeds necessary for its dismantling: addressing the human-canine pair as one subject of discipline, it fostered the strengthening of the bond between them, producing responses of empathy with the dog's experience in the human.

4 The Soul of a Trainer: Crossover Trainers, 1980s–2000s

In order to document the discomfort experienced by many trainers using the classic disciplinary methods – and running the risk of muddling the chronology of events – I now need to turn to books published in the 1990s and early 2000s to discuss the concept of "the crossover trainer." Books by these trainers – Morgan Spector's *Clicker Training for Obedience* (1999), Pat Miller's *The Power of Positive Dog Training* (2001), Vicki Ronchette's *Positive Training for Show Dogs: Building a Relationship for Success* (2007), Gail Fisher's *The Thinking Dog: Crossover to Clicker Training* (2009) – written often several decades after the author's first exposure to discipline-oriented obedience dog training classes, focus on experiences of dissatisfaction with what the writers termed "traditional training." Such accounts often form the introductory chapters of books that advocate reinforcement-based methods, which begin to proliferate in the early 2000s. In fact, the personal account of how the author found his or her road to positive training becomes such a staple of turn-of-the-century training manuals that it spurs the development of a new term: the crossover trainer; the trainer who switches training practices from traditional to reinforcement-based training as a result of experiencing discomfort with the old techniques.

These accounts are all characterized by a sense of emotional urgency, a zeal that is almost religious in its intensity and an equally quasi-religious desire to "convert" the readers of the book to the style of training the author is advocating. Morgan Spector, in a chapter aptly titled "A Personal Note: The Relationship – Or, Where This All Comes from and Where It Can Get You," describes his first encounter with clicker training at a seminar with Karen Pryor and Gary Wilkes in 1993 as "a conversion experience" (Spector 1999, 2). Pat Miller writes of "an epiphany" (Miller 2001, 8). This urgency reflects the raw emotions that accompanied the entry of reinforcement-based training onto the dog training scene in the 1990s and the intensity of the debate surrounding it. It also points to the fact that it seemed impossible, at the time, to even attempt to amalgamate the old with the new: once a trainer crossed over, it was impossible to go back. It did not just seem impossible, it really was: the trainers formed two opposing camps that created two distinct social worlds, shunning members of the opposite camp. Each camp defined itself as adhering to a philosophy that knew and advocated "the truth" about the human-canine relationship. Reinforcement-based training was not just a technology of training but – in a Foucauldian vocabulary – it initiated an epistemic shift in the discursive field of dog training. It opened the field of training to the possibility of thinking in completely new ways.

Yet all of these accounts begin with a sense of discomfort that is vaguely ethical. Something is off, but it is, at first, difficult for the writers to define the problem. As the trainers narrate their personal stories, it becomes clear that the dissatisfaction invariably results from a certain incompatibility of the practice they are engaging in and the way that they define their bond with their animals. These writers speak of love between the human and the dog in a way that no trainers before them had, and they view the training practices they encountered as they entered the field as being at odds with the relationship they desired to have with their dogs. The narratives they present share similar features: they begin with the purchase or adoption of a dog, the recognition of the need to influence the dog's behavior, the grudging and somewhat unwilling participation in obedience classes; the gradual immersion in the world of dogs, accompanied with a growing sense of dissatisfaction with the practices of dog training. In *Clicker Training for Obedience*, Spector recalls his story of buying Woody, a sheltie, and training him using "traditional 'J&P' ('jerk and praise,' sometimes referred to as 'jerk and puke') methods" (Spector 1999, 1). Spector's two dogs, Woody and then his younger sheltie Sam, suffer health-related issues that Spector links, *post factum* of course, to the training methods he employed. He speaks of Sam's "depressive episodes," of Woody's big heart in his fight with the illness that overtook the little dog and of his own overwhelming love for the dogs. It is not his lack of objective success in the training ring that leads Spector to cross over to clicker training; it is the feeling of love that he finds incompatible with the leash jerks. He trains because he believes he can achieve "a feeling of community and 'family' " not because of ribbons, and he is willing to admit his fault in using methods that do not allow him to achieve that goal. Similarly, Pat Miller explains her decision to change her training techniques as stemming from the rift in the relationship she had with her terrier mix, Josie, caused by the application of the ear pinch in teaching Josie to retrieve. One day, as Miller was getting out the training equipment for the next training session for Josie's Utility Dog title, she saw the terrier hide under the deck and refuse to come out. Miller writes: "Josie's behavior was a rude awakening to me. What was I doing to my dog? Were a few scraps of satin ribbon and a certificate to hang on the wall worth the damage I was doing to Josie's psyche, our relationship, and my soul?" (Miller 2001, 9).

While the gradual gentling of methods coupled with the desire to train in a more humane way had been discernible in dog training since the 1940s, what happens in the 1990s is something more than an extension of this evolutionary shift. It is, for the trainers, a sudden and abrupt switch: an epiphany, a conversion experience. The use of reinforcement is technically nothing new; it had been a part of dog training for a long time, as the chapter on nineteenth-century trainers proves. Still, what the diaries of the crossover trainers demonstrate is a sudden epistemic shift. Once the trainers accept the methodology of reinforcement-based training, a whole new way of looking at the human-canine relationship becomes possible. At the same time, the acceptance of this new paradigm provides a discourse for speaking about that which remained unspoken in the old one, yet which was somehow sensed, experienced if not verbalized. This is particularly evident in *The Thinking Dog* by Gail Fisher. Fisher's literary output in the form of numerous training manuals reflects the path of gentling experienced by so many trainers. Yet, she still speaks of the moment of understanding what reinforcement-training was about as an "A-ha!" moment: "this about-face, this complete change of attitude happened when I became a crossover trainer" (Fisher 2009, x). That which had been previously considered a problem, no longer is because the epistemological paradigm for conceptualizing human-canine interaction changes completely, resulting in feelings of exhilaration and liberation. It suddenly becomes possible to conceptualize training interactions without punishment.

5 Off the Leash

The liberational aspect of reinforcement-based training is presented in the crossover memoirs via a potent visual metaphor: the taking off of the leash and the collar; resulting in freedom from the constraints of the training apparatus of traditional training, which I have described as the tools of disciplinary surveillance. In all of the crossover accounts under consideration, the leash and the training collar play a significant role in eliciting the discomfort of the trainers. They very literally leave their marks on the bodies of the dogs. Spector describes how his wife noticed that the fur around Woody's neck was gone and comments: "I had unwittingly shaved him to Naval Regulations with all the 'sawing' of the choke collar" (1). Ronchette describes her ambivalent experience with the prong collar:

> After the first night of class, my hands were raw from Boris pulling on the lead while I struggled to hold onto him with it. On the way home I cried and thought, "I can't do this. This is too hard." I felt better the next day and went to class again. Although I was uncomfortable with the recommendation to use a prong collar on Boris, I did. I was assured it would help, and indeed it did. It made all the difference in the world: power steering. We were taught to pop the leash when the dog was out of position and give praise and hot dog pieces when he did something correctly. (Ronchette 2007, 6)

Sarah Stremming, the author of *The Cognitive Canine* blog, thus describes her experiences of using the prong collar:

> I pulled the prong collar out of the bag, and turned to Kelso. His ears flattened. He lowered his head. He licked his lips and waited for me to attach it around his neck. I saw him, and I wish I could say I put that archaic tool back in the bag, but I didn't. I put it on him and we proceeded to join the group heeling exercise in class. That moment, and many others like it, haunt me still. (Stremming 2017)

The overwhelming sense of guilt elicited by the use of the training apparatus is juxtaposed with the sense of relief experienced after abandoning the use of these "instruments of torture." Miller writes of using toys and treats instead of collars and ear pinches in the training of Josie: "when I get out her training toys, her eyes sparkle and she dashes in happy circles around the yard" (Miller 2001, 9). Morgan Spector writes of the first decision he makes after attending the life-changing Karen Pryor/Gary Wilkes seminar: "My first step, simply, was to throw away the leash and the training collar. It was at that point, I think, that I started to become a trainer" (Spector 1999, 2). Taking off the leash becomes a metaphor of freedom: for the dog, but also for the trainer who is no longer limited by the constraints of a methodology that relies on the human assuming the role of the strict disciplinarian who administers corrections. Using her trademark technique of defamiliarizing training by assuming the dog's perspective Jean Donaldson writes in *The Culture Clash*: "Entire methods of training are founded on specific types of collars. It must seem like bad science-fiction to dogs that their owners' choice of training methods was, until quite recently, a choice of hardware around their necks" (Donaldson 1996, 36). One of the first trainers to attempt off-leash training in a group context was Ian Dunbar, founder of the American Association of Pet Dog Trainers, who also popularized the idea of puppy classes in the 1980s. In an interview with the author of *One Nation Under Dog*, Dunbar recalls: "Puppy training was, you bring your pet to class, you let him off leash, and you teach him off leash. (....) The rationale was, he's off leash at home ... And it was fun. It was so different from having the dog on a leash, pulling and pushing" (Dunbar quoted in Schaffer, 2009, 191).

It is worth recalling that William Koehler defined the goal of training as complete off-leash reliability. In fact, off-leash reliability was something of a categorical imperative for Koehler: nothing that fell short of that goal would satisfy him.[4] The leash shaped the body into docility, but this docility later enabled the dog a certain amount of freedom that he would not have experienced otherwise. Somewhat ironically, as the leash becomes obsolete in the process of training, it also becomes a constant element of the dog's life outside of the home and the training session. It is in the 1990s when leash laws begin

---

4 Interestingly, it seems that this categorical insistence on obedience has made William Koehler a favorite trainer among philosophers. Not only did he influence the writings of Vicki Hearne, as discussed in Chapter 6, he is also used by Randall Havas to explain Nietzsche's concept of reliability in the article "Individuality as Reliability: A Dog Trainer's Guide to Nietzsche" (2012). Unlike Hearne, Havas does not advocate Koehler's methods, but he sees them as practical applications of Nietzsche's understanding of reliability and its tie to individuality. From the more practical side of the Koehlers' training business, this insistence on off-leash reliability as the ultimate goal of training has required contorted explanations for today's world, where if a dog is off-leash in public space, his owner is likely to get a fine. Still, the Koehler website, which is where the brand (owned by a former student of Koehler) is mostly active nowadays, continues to argue for off-leash reliability "because leashes break. Because dogs will slip out of collars and harnesses. Because doors and gates will be left open. Because fences will be jumped over, dug under, or rammed through" (Koehlerdogtraining.com 2017).

to be energetically enforced in the US (Walsh 2011, 27–34) and when it all but becomes illegal to let a dog run free in any public areas. The explanation of the necessity of obedience training for pet dogs through the framework of proving dog's use value to man was strained right from the beginning; it barely made sense as obedience competitions entered the scene in the 1930s, even despite trainers' attempts to naturalize that line of reasoning. What did make sense was the potential of using training to minimize the harm that could potentially be caused by dogs, although, as this narrative unfolds, it becomes clear that simply ensuring the dog's physical connection to his owner through the leash is a simpler and less resource-intensive way of achieving this goal than a public investment in the training of pet dogs. Consequently, the zeal of dog training as a social duty, so explicit in Saunders's stories of churches and libraries opening their doors to dog obedience classes dies down, and dog training changes shape again. It becomes privatized in more ways than one: not only does it move away from the sphere of interests of public authorities and turns into a business enterprise, it also becomes explicitly a tool for the development and strengthening of the individual human-canine relationship rather than a tool of population management.

In all of this, Spector's words about becoming a trainer upon throwing away the leash also touch on the major redefinition of training that the shift from "traditional" to "clicker" training entailed: it is a shift away from disciplinary control, not a shift away from control (even if that is what Spector imagined) but towards a different type of control. The maintenance of disciplinary control required the tools of discipline to operate somewhere in the background, to recall Saunders's advice about what to do if the dog disobeys while off-lead (put the leash back on and remind the dog of its presence). Paradoxically, thus, the doing away with the tools of disciplinary control is a sign of control, but of a different kind of control: of affective control. It is the move away from discipline that makes it possible to speak of love and bonding. Reinforcement-based training becomes a technology for putting into practice the rule of love.

6 Feeling Power and Positive Dog Training

While Foucault's major thesis posits that it is more productive to read changes in modalities of power as tied to increased efficiency – and not increased humanitarianism – he does not disregard the power of affect and the existence of these feelings of discomfort, as I have termed them, does not counter the thrust of Foucault's analysis of power relations. On the contrary, his analysis, particularly in the later stage of his life, is more multi-dimensional and allows for the recognition of the multi-nodal relationship between affect and power. Even if Foucault has not been the most influential thinker for the affective turn, he did recognize affect's potential for destabilizing institutional arrangements. Lauren Gilmette, arguing for the relevance of Foucault's late work for affect theory, notices that while Foucault does write about how affective responses have themselves been defined and normalized by networks of power, he also notices that "affect also exceeds this range, opening space for resistance when one feels ill at ease with relational and/or institutional arrangements" (Gilmette 2014, 286). Foucault's definition of genealogy from "Nietzsche, Genealogy, History," recalled in the epigraph to the introductory chapter of this book, speaks of "sentiments, love, conscience, instincts" as the "unpromising places" on which genealogy must focus in order to "isolate the different scenes where they engaged in different roles" (Foucault 1984a, 76).

As the editors of the special issue of *Parrhesia* on "the affective turn" argue, "for Foucault, far from being a mere descriptor of emotional states, affect is the site of the production of the modern soul" (LaCaze and Martyn Lloyd 2011, 1). Thus, using Foucault to look at the affects involved in the turn away from the application of classic disciplinary techniques in dog training is quite productive. It is the trainer's affective response to her dog's behavior during a training session that causes Pat Miller to exclaim that she has "ruined her soul." The experience of discipline creates in this trainer a wish to shape her training relationship with her dog differently than she had been taught to. In the interview "On the Genealogy of Ethics" Foucault posits that in contemporary society it is feelings that are the most relevant for morality (1984b, 352). In other words, it is the feeling of ethical discomfort, the subjective experience of feeling somehow off, that leads to guilt. What generates the experience of discomfort is the physical interaction with the dog's body in the process of training. It seems that what can be identified in the accounts of the crossover trainers are crucial aspects of Massumi's influential definition of affect, provided in the introduction to the translation of Deleuze and Guattari's *A Thousand Plateaus*; that is of affect as the result of encounters with other bodies that increase or diminish the body's power of activity (Massumi 1987, xvi). Even though for Massumi affect is to be distinguished from feeling and emotion, a link can be posited between Foucault's feelings and affect. What makes these "feelings" relevant for morality is that the subject can act in accordance with them; search for other ways of relating to the dog; ways that would diminish the discomfort the human is experiencing as a result of empathy with the canine subject of training.

In a 2013 essay Nicole Shukin coins the concept of "feeling power," which is not in itself a Foucauldian term, but originates from a Foucauldian genealogy of power. Shukin finds the term "feeling power" useful for conceptualizing the

"rule of love which begins to govern human-animal relationships in the twentieth century" (Shukin 2013, 187). Feeling power posits that dogs are capable of acting out of loving attachment for their humans. Shukin shows how feeling power operates in the perception of military canines, whose acts of bravery are seen as stemming from their love for their guardians (and, by extension, also for the nation), just as it can be seen in the institution of Welcome Dogs who greet returning soldiers and help them transition into civilian life through the offering of unconditional love. Conceptualizing dogs through feeling power results in seeing them as full-fledged subjects of governmentality whose capacity for unconditional love and attachment is cultivated as an instrument of security (Shukin 2013, 178). In other words, the fact that we fantasize about how loving dogs are can result, and has resulted, in their incorporation into the project of governmentality and – as I argue in the next chapter – in investing them with a potential for self-governmentality. Feeling power is a fantasy but one which has implications for real-world interactions with animals: it makes us likely to long for interactions that confirm this fantasy. Shukin analyzes feeling power on the literary example of Jack London's *White Fang* (1906), where the concept structures the feel-good plot in which Buck finally meets Weedon Scott; the first person to stroke him, instead of striking.

When it comes to the actual technologies of animal training, feeling power seems to account for the desire for training practices to express the "rule of love;" so evident in the memoirs of the crossover trainers. This desire can be glimpsed already in the nineteenth century with the declarations of gentleness (that seem to contradict the actual use of the whip); it is also present in Saunders's admonition to never kick a dog and in the "wincers' " refusal to accept Koehler's helicopter corrections. Even as the desire to base training in feeling power increases, the practical techniques of dog training do not keep up with the rhetoric of gentleness. A fantasy is one thing, but the practical aspects of human-canine interaction reveal feeling power to be just a fantasy: for over a hundred years of the existence of training manuals it proved to be impossible to conceptualize a training technology that included only stroking and not striking. A perfect example of this discrepancy between the fantasy of training as an expression of love and the reality of training techniques is the story of William Koehler's participation in the filming of Walt Disney's animal movies; most explicitly, *Big Red*. The plot of the 1962 movie, based on a 1945 book by Jim Kjelgaard, follows the story of an Irish setter, whose cruel trainer expects unconditional obedience from the dog, treating Red as an income-generating commodity. An orphaned boy, through love, devotion and kindness, gains the respect of the dog who, in an act of selfless gratitude, ends up saving the boy's life. In other words, this is a story of how the rule of love triumphs over the harsh disciplinary techniques employed by the cruel trainer: a story that seemed compatible with the rise of feeling power in the twentieth century, as described by Shukin. However, the canine actor in the movie was trained by William Koehler using "certified techniques by movieland's most experienced dog trainer," that is through Koehler's extreme version of "jerk and praise." What this means is that the story that condemns harsh training techniques was created using the very techniques that it opposed. Feeling power was, and still is, a part of the fantasy of dogs' unconditional love for humans. Koehler attacked the notion of unconditional love as a fantasy in his writings, even though he perpetuated it through his involvement in Disney's feel-good movies. Yet the growing importance of feeling power in relation to animals clashed deeply with the actual experience of training, as evidenced in the accounts by crossover trainers. As the discomfort they experienced increased, the change-oriented trainers became more willing to sacrifice the fantasy of unconditional love for a technology of conditioning love.

Among trainers there existed recognition that love in itself is not a technology, and that there needs to come into being a technology of training that is compatible with their love for the dog and one that produces reciprocity in the dog. The arrival of reinforcement-based training promised to provide just that: step by step instructions for producing love. Spector writes: " 'Mom's (or dad's love)' isn't enough. The dog will (or should) get that anyway. There has to be something intrinsic in the work to create the necessary motivation and consistency. That's what positive reinforcement is all about" (Spector 1999, 12). Reinforcement-based training is a technology of affect in the purest sense: not only is it an expression of the guardian's love but its goal is to influence the dog's affects: to make the dog love the handler more. The fact that dad is bracketed in the quotation from Spector is significant. Emotional labor, the work of care, had long been associated with women. As training becomes redefined as the technology of feeling power, it also becomes much more strongly the province of women. At the time when Spector was writing his book, men in competitive obedience were already outnumbered by women. As the sport of canine agility emerged in the 1990s, the prevalence of women was overwhelming and is well documented by social scientists and confirmed by more auto-ethnographic accounts emerging from dogland (Haraway 2008; Lund 2014; Hughes 2015). While Walker and Saunders did not enter the field of dog training with the goal of changing the gender balance among trainers, the competitive sport they initiated is now almost completely taken over by women.

The confluence of the two types of grounding of this turn to reinforcement-based training: in feeling power, as described here, and in efficiency, as discussed in the next chapter, prove that the two shifts – that towards a more loving relationship with the dog and that towards a more efficient way of governing others – are intertwined in a way that should make us uneasy. It is not coincidental that the authors discussed above argue that their way of training is not just more humane but also more effective; "Clicker is quicker!" was the motto of reinforcement-based trainers at the time. Even though their disagreement with the traditional methods is not based on their ineffectiveness but on the emotional discomfort they elicit in the trainers who experience a sense of intersubjective empathy for their dogs, they still emphasize that what they are advocating is a training methodology that is both humane and effective. In fact, it is superior in effectiveness to traditional training. Fisher writes of clicker training: "It is not a New Age touchy-feely gimmick, as those who try to marginalize it may claim" (Fisher 2009, 10). Spector also encourages the reader to take up reinforcement-based training because it is a methodology that guarantees results: "[o]perant conditioning will produce a dog that it more focused on his or her work, and is therefore more attentive in the ring. It will produce a dog that works with more precision because he or she understands clearly what you want" (Spector 1999, 2). The simultaneous use of the vocabulary of love and human-canine kinship and the vocabulary of production and efficiency seems to be a paradox that can only be explained through a closer contextual analysis of the methodology these trainers were embracing as the technology of love: B.F. Skinner's operant conditioning.

CHAPTER 5

Power without Coercion: From Governmentality to Self-Governmentality, from Discipline to Self-Control, 1984–2000s

> There is now a sort of shift: these games of truth no longer are concerned with coercive practices but with the practices of self-formation of the subject.
>
> INTERVIEW WITH MICHEL FOUCAULT, ON JAN. 20, 1984, "THE ETHIC OF CARE FOR THE SELF AS A PRACTICE OF FREEDOM"

•••

> What could we do? They mentioned in several classes that they thought we should get into dog training. We bought a bunch of dogs. We knew there were so many dogs in the country, and people always wanted to get them trained. So we thought this would be a cinch. We'd tell people about this new, humane way of training and they'd be talking to us by the thousands. Nobody listened to us. The dog food companies, etc., didn't think it was all that new. People had been training dogs for thousands of years. That's the kind of greeting we got.
>
> UNDATED INTERVIEW WITH MARIAN BRELAND BAILEY BY SOPHIA YIN

∵

It has rarely been explored why behaviorism made its way into dog training only in the 1990s. The few sketchy accounts of the history of dog training speak of the entry of science into the age-old practice (Burch and Bailey 1999; Lindsay 2001): a kind of "Enlightenment" for the murky field steeped in superstition. "When the student is ready the teacher will appear" was a popular quote (attributed – not coincidentally – to Zen philosophy) among dog trainers in the 1990s, as recorded by Gail Fisher in the introduction to *The Thinking Dog* (Fisher 2009, vii). What the context of the use of this quote implies is that a certain mental change had to take place among dog trainers – a change consisting

 | DOI:10.1163/9789004380295_007

mostly of experiencing training as an activity with the potential for human-canine relationship-building outside of a blatantly coercive framework – that led them to look for more ethical ways of interacting with dogs. The broader relationship with pets had to become redefined in ways that have been extensively theorized by the recent wave of animal studies scholarship but that can basically be summed up through the growing importance of humans' emotional connection to pets, whose status rises to that of significant others and family members. It is the rise of "feeling power" outside of the sphere of dog training that generates a desire for change within dog training. It is not coincidental that the desire to train dogs without the use of physical punishment and their concurrent rise in conceptual status to kin, mirrors changes in parenting and educational practices. In *Governing the Soul*, Nikolas Rose writes of the introduction of "contingency management" in institutional educational regimes in order to replace corporal punishment (Rose 1999b, 241). From this perspective, it is not surprising that the publication of a child-rearing manual that included the word "train" in the title – the conservative homeschoolers Michael and Debi Pearl's *To Train Up a Child* – caused a public uproar in 1994 (Orłowska 2016, 38). The disciplinary understanding of training, as establishing behavior patterns using punitive techniques, was seen as obsolete in the world of contemporary parenting. But this definition was also becoming obsolete in the world of dog training, which was searching for a methodology that would allow a makeover of this disciplinary field. It was the broader changes in the human-animal relationships in modern society and the rising recognition of the role of affect in interpersonal relationships that forced dog trainers to adjust and come up with methods that reflected these qualitative changes.

Clearly, at the time when Keller and Marian Breland, former students of B.F. Skinner, attempted to conquer the dog training world with their new training methods in the 1940s, the changes had not yet taken place, but when Karen Pryor's groundbreaking book *Don't Shoot the Dog!* came out in 1984 it made a huge impact, even though it was not intended to be read as a dog training manual; the dog in the title is a metaphorical representation of dealing with unwanted behaviors. In fact, Pryor violently objected to the title, worried it could be misleading (Pryor 2009, 8). In accordance with Skinner's spirit of universalism, Pryor published a book that explained how the principles of operant conditioning can be used to modify the behavior of any organism without resorting to physical punishment. The publication of what turned out to be a very aptly-titled bestseller altered the world of dog training and the author's life: Pryor made a career change from working with marine mammals to working with dogs and (absolutely not coincidentally) running a successful business venture. Simultaneously, behaviorism became the new science of dog training. If there had been an "old science" – a thesis some would most certainly disagree with, claiming that dog training did not seek to validate itself through science – it would have been ethology, but by the 1980s ethology, at least as it related to dogs, had become morally corrupt for "the wincers" as a result of its close association with traditional training techniques. It seemed as if the wolf experiments of the 1940s and the claims made about dogs in *Man Meets Dog* (1953) and *King Solomon's Ring* (1952) by Konrad Lorenz – who had won the Nobel Prize, after all – corresponded with methods based on physically asserting human dominance. Ethology, at its state of development in the 1980s, could not serve as a tool of change. Still, it is by no means obvious that behaviorism – which by the 1980s was becoming displaced by cognitivism within mainstream psychology – would logically follow as the science of dog training; even if it had become omnipresent in the training of non-companion animals, as the story of the Brelands illustrates. Discouraged from dog training by the cold welcome they received, the Brelands established a business, Animal Behavior Enterprises, which trained numerous non-companion species to perform at county fairs and star in commercials. The question remains: why was behaviorism deemed the methodology that could fulfill these mostly emotional needs that arose among trainers (also pet owners, more generally) in the last decades of the twentieth century? How could behaviorism become a technology for turning the fantasy of "feeling power" into a coherent training methodology? I propose that reading Skinner alongside Foucault can help in answering this question.

Reading the two thinkers alongside one another makes it possible to answer the question of why Skinner has been so influential for the development of positive dog training: a fact so obvious it is hardly ever questioned. Yet, I have argued that the desire to revolutionize dog training came from a strengthening of the emotional bond between the human and the canine, which, while related to multiple factors, included these two concurrent ones: the firm embedding of the human-canine dyad within the framework of governmentality and the human half of the dyad's dissatisfaction with the first-hand experience of the disciplinary techniques of obedience classes. Consequently, Skinner seems an unlikely influence for trainers who argue for change of training practices on the basis of love and bonding, as Morgan Spector and Gail Fisher did. Skinner's emotional connection to dogs was very weak, to say the least. To begin with, B.F. Skinner was never interested in dog training. While instances of him training specific dogs to perform a particular trick have been recorded – and in 1952 he even participated in a photo session for *Look* magazine which dared him to train a dog to move her head within 20 minutes (*Look Magazine* 1952) – he was not a dog person. His laboratory animals were first rats, then pigeons and in his later years, as he moved out of the laboratory, Skinner became increasingly interested in applying the principles of operant conditioning to humans; though his first attempt came already during his laboratory stage and consisted of a novel, titled *Walden Two* (1948), which described a utopian society organized according to behaviorist principles. He was not interested in species specificity but in a universal account of the similarities of contingency management across species: a fact repeatedly pointed out in so many critiques of the application of Skinner's ideas to dog training.

Secondly, Skinner did not seem particularly concerned with animal welfare. While he did believe in the power of positive reinforcement, he was primarily an experimental scientist who tested claims in the laboratory and promoted those that provided him with the best results. Had Skinner not believed that he had empirically proved the effectiveness of positive reinforcement, he would not have promoted it as the preferred tool for shaping behavior. He also used positive punishment in the laboratory experiments that led to the claims in favor of positive reinforcement posited in *The Behavior of Organisms* (1938). The Skinner box, after all, not only delivered food but also electric shocks and the sterility of this environment in no way fostered the strengthening of the bond between the trainer/scientist and the animal in training. In fact, the scientist's role was limited to recording the animal's behavior. Project Pelican, which Skinner initiated during World War II, consisted of training pigeons to become suicide bombers: guiding missiles to their targets. The project, inspired by Japanese kamikaze attacks (Capshew 1996, 134), was never carried out but clearly Skinner had no ethical qualms about sacrificing the lives of the animals he was training. In the 1970s, after the publication of his bestselling *Beyond Freedom and Dignity* (1971), Skinner was generally perceived as a manipulator and even a fascist – which he most certainly was not – and the idea of scientifically controlling human behavior was seen as a dystopian vision, akin to those imagined by fiction writers like Aldous Huxley or Anthony Burgess.

I would venture to say that the reason for the delay in adopting radical behaviorism by dog training practitioners – which took place only in the 1990s and early 2000s, that is fifty years after Skinner's first publications about the effectiveness of positive reinforcement – should not be explained away, as it often is, by the trainers' ignorance of science. The real reason behind the slow and somewhat grudging adoption of behaviorism is an incompatibility between the project dog trainers felt they were engaged in and what they believed behaviorism was delivering. There was a dissonance between the goals of the two fields. The trainers were looking for a more ethical way of relating to dogs; to them, Skinner's behaviorism seemed anything but ethical. Those qualms are analyzed in chapter 6, while this one tries to account for the eventual convergence of behaviorism and dog training. I argue that Skinner's technologies of behavior shaping spoke directly to ideas that appeared slightly later in completely different disciplinary fields and led to the fall of what has historically been defined as "the autonomous subject;" a term Skinner actually uses judiciously and critically throughout *Beyond Freedom and Dignity*. Skinner's famous universalism – the claim that all organisms learn as a result of the same principles – contributed to the blurring of the human-animal divide. The application of Skinner's principles to the training of dogs offered to extend subjectivity to dogs and posited that canines are capable of self-governmentality; a feature that had traditionally served as a marker of human-animal difference.

1 Had Foucault Read Skinner?

Clearly, the personal paths of these two men never crossed. Their disciplinary background and intellectual heritage differed. The two thinkers have rarely been read alongside one another, mostly because their audiences have been so different. Yet, Foucault does show an awareness of Skinner, and Skinner's name is mentioned in the 1978/79 College de France lecture series, *The Birth of Biopolitics* (English translation: 2008). Foucault casually drops Skinner's name when discussing the specificity of American neo-liberalism, immediately after recalling economist Gary Becker's concept of *homo oeconomics*: a key figure of the neo-liberal marketplace; a person whose conduct is sensitive to modifications of the environment and who responds to these modifications in a non-random way (Foucault 2008, 269). Foucault goes on to say that once we accept that the conduct of *homo oeconomicus* is governable, it is possible to come up with "a set of techniques" that make it possible to influence his behavior:

> You find these methods in their purest, most rigorous, strictest or aberrant forms, as you wish, in Skinner, and precisely they do not consist in analyzing the meaning of different kinds of conduct, but simply in seeing how, through mechanisms of reinforcement, a given play of stimuli entail responses whose systematic nature can be observed and on the basis of which other variables of behavior can be introduced. (270)

It is significant that Skinner's name comes up in a text that is concerned with neo-liberal economy and the techniques of "American behaviorism" – Foucault really does emphasize the American specificity here and adds that there exists little literature in France on the topic – are tied up with the marketplace. In fact, Skinner's techniques are used to reveal what Foucault dubs the paradox of neo-liberalism: that the individual player on the market is seen as both free and governable at the same time. He is free, and he needs to be left alone in the way of *laisser-faire*; his freedom should not to be restricted. Yet, on the other hand, because he acts in a rational way, he is also always influenced by environmental conditions. His behavior can also be predicted and managed via contingency management. In other words, Foucault points to the problems with the definition of freedom within neo-liberalism; the kinds of problems that have by now become prominent in critiques of neo-liberalism: neo-liberalism tricks the subject into believing he is free to shape his life, while in reality the conditions of this freedom are restricted via the existing institutional framework; consequently, the subject's failure to succeed is blamed on the subject's personal inadequacy and not on the shortcomings of institutional conditions of possibility. Of course, most of Foucault's work had explored how the subject was formed through power and his comments on American neo-liberalism are no exception. What makes them pertinent for this discussion is that Foucault reads Skinner's behaviorism as the primary set of techniques for the shaping of the neoliberal subject; coincidentally, capturing the promise but also predicting the intellectual downfall of behaviorism in the world of dog training. Behaviorism offered a semblance of neo-liberal subjectivity to the dog and this is what the trainers who wanted to respect the emotional bond between the human and the canine wanted, or at least thought they wanted before the shortcomings of the notion of neo-liberal subjectivity became apparent. It did matter that the technologies were not coercive, but it seems more significant that this promise of freedom from coercion bled onto other fields: freedom to actively shape the training process and the concurrent recognition of the dog's intellectual capacity to act in a rational way, at least to some extent. In acknowledging the dog's capacity for rational behavior, the trainers saw validation of the emotional connection between themselves and their dogs as well as appreciation of the dog's significant place in their lives.

Adopting the framework of protocols of reinforcement also made it possible to do away with the paradigm of obedience/dominance/submission that trainers found compatible with harsh training methods that they vehemently opposed. It enabled the death of obedience and the birth of "the thinking dog": the dog who could choose wisely from many options. The disassociation of obedience (seen as the province of training) and intelligence (often perceived as the enemy of training) had been a basic assumption of dog training since the publication of Most's *Dog Training: A Manual*. This belief changes with the turn to behaviorism, although it seems a counterintuitive change because Skinner claimed any organism with a brain stem was capable of being trained. For the behaviorist, an animal's superior intelligence is also not a predictor of training success. Yet training a dog using Skinner's methods really made the human perceive the dog as a highly intelligent being. Karen Pryor writes in her personal recollections of popularizing clicker training:

> I take particular pleasure in the people who have tried this kind of training with their dogs (...). Their first reaction, as the dog begins to offer them behaviors, is a thrill of personal success: wow, finally, look, he did what I wanted. And the second reaction is from the heart: I had no idea I had such a smart dog. I used to think he was such a nuisance. I *love* this great dog. (K. Pryor 2009, 13)

Pryor's observations reveal how the technology of behavior shaping is also a technology of affect: and a two-way technology at that. It is obvious that it can serve to increase the dog's desire to be with and interact with the human. What is less obvious but equally important is that the human's behavior upon training the dog results – as Pryor claims – in a greater love for the dog. It is the dog's knowability, the predictability of his reactions that behaviorism promises that contributes to this effect, but the perception of the dog as engaging in problem solving is the key. The dog is perceived as practicing thinking skills as he engages in training interactions. There is a deep irony in this because Skinner's behaviorism most certainly did not claim that animals were capable of abstract reasoning. Thinking was simply outside of Skinner's sphere of interests. Many of the early reinforcement-based trainers were adamant that the attribution of intelligence to dogs has caused them much harm and that the path towards a loving relationship necessarily included a difficult renouncement of canine intelligence. In what was potentially the most influential dog training book of the decade of the 1990s, *The Culture Clash* (1996), Jean Donaldson criticized what she termed as the fallout of the Disney myth: the attribution of intelligence leads to unrealistic expectations and, in turn, to unfair treatment of the dog. Donaldson complained: "We are obviously fascinated by the notion that dogs might, just might, be really, really, smart. Why do we keep referring to this particular yardstick, intelligence, to evaluate another species?" (12) Still, the actual experience of training that utilized operant conditioning inevitably brought about results like the ones observed by Pryor: "I had no idea I had such a smart dog."

The behaviorist assumption that all organisms, regardless of species, are subject to the same principles of conditioning can also be read as a step towards the blurring of the human-canine divide and, as such, it did seem appealing to the change-oriented trainers. It is not surprising that the dog training revolution started with a book that was not about dog training. Karen Pryor's *Don't Shoot the Dog* includes behaviorist solutions to problem behaviors ranging from addiction, depression, anxiety and the temper tantrums of toddlers to teaching dogs to play Frisbee and dolphins to jump through hoops. The basic premise that we are all the same, that is we are all organisms subject to the rules of operant conditioning, seems to have further fed the interspecies empathy awakened by the experience of traditional training. Reflecting on the hostility to behaviorism among some people, Pryor comments: "It's even more ruffling to think that we, like laboratory rats, might be controlled by mysterious systems of rewards and punishments, against our will and without our knowledge, as behaviorism seems to imply" (K. Pryor 2009, 9). Change-oriented trainers did not want to hear that one can correct a dog simply because the dog is a dog and not a human being. The equality posited by behaviorism between different species actually raised – rather than lowered – the trainers' definition of what constitutes fair treatment.

It is also significant that, according to Foucault's *The Birth of Biopolitics*, in order for the concept of *homo oeconomicus* to come into being, that is in order for someone's behavior to be considered rational, the threat of physical pain needs to be removed from the list of potential consequences of one's choices. Foucault speaks of this as "an irreducible": "The painful or non-painful nature of the thing is in itself a reason for the choice beyond which you cannot go. The choice between painful and non-painful is a sort of irreducible that does not refer to any judgment, reasoning, or calculation. It is a sort of regressive end point in the analysis" (Foucault 2008, 272). In other words, we cannot speak of rational activity if the choice is made between pain and avoidance of pain. Pain needs to be taken out of the equation in order for rational (if not completely free) decision-making. Skinner-influenced reinforcement-based training truly takes this principle to heart: the choice between receiving a correction and not receiving a correction (this is the choice from traditional training) is not considered a real choice. In a way, for the rhetoric of choice to even begin to operate, the threat of physical pain needs to be eliminated from the potential consequences of behavior. This is the dogma of positive training; a dogma because it is clear that a dog trained using corrections is also making certain choices: the choice to avoid the correction being the most obvious one. Yet, for positive trainers this choice does not count as "choice proper." This understanding of choice comes from a certain philosophical tradition, which can be traced back to David Hume, and which becomes a guiding principle of the contemporary incarnation of biopower, which Foucault sees as linked to the rise of liberal capitalism.

Additionally, the actual practice of training via behaviorist techniques inevitably includes moments of the dog actively trying to figure out what it is that he is being rewarded for: offering behaviors in order to test which one is being reinforced, an activity that is consistently labeled as "thinking" by those who have experienced this kind of interaction. While trainers do recognize that succeeding at a training task taught through shaping does not actually require thinking,[1] this does not seem to lessen their enthusiasm for using this word, which seems significant enough to make it into the descriptions of many of the training techniques and even into the title of Gail Fisher's book about clicker training: *The Thinking Dog*. If we return to Foucault's reading of Kant in "What is Enlightenment?," Foucault's words on the relationship between obedience and reason in Kant gain a special poignancy in light of this emphasis on the necessity of teaching the dog to think expressed by clicker trainers. Foucault summed up Kant's ideas by recalling Kant's proposition of the contract with the despot in which "the public and free use of autonomous reason will be the best guarantee of obedience" (Foucault 1984d, 37). I had initially used Foucault's recollection of Kant's distinction between obedience and reason to conceptualize Most's distinction between a dog's intelligence (or lack thereof) and trainability; a notion that broke with the nineteenth-century idea of the sagacity of the trained dog. Yet, it is obvious that for Most this distinction was significant only because he argued that one does not need to the capacity to reason in order to be obedient. A seismic shift takes place with the entry of clicker training onto the scene: the ability to think begins to be seen as (potentially and preferably) connected to "obedience." The *homo oeconomicus* canine, the dog who is a rational player, can figure out that it is advantageous for him to behave in a specific way: his rationality leads to obedience (except that compliance with the handler's wishes ceases to be defined as obedience within reinforcement-based training). The conclusion that rationality results in obedience would actually constitute a huge misreading of Kant's idea from "What is Enlightenment?" and yet it seems to be the misreading that enabled this turn in thinking about dog training, and – of course – the bigger framework

1 In *Don't Shoot the Dog* Karen Pryor proposes a game called The Training Game, in which humans trade places with their animals as subjects of training through the technique of shaping. The game, which truly was a party pleaser among animal trainers for a long time, consists of the "trainer" marking subsequent approximations of the desired behavior with some kind of a marker (whistle, clicker or verbal marker). Pryor writes: "The subject gets to discover that in this form of learning, brains don't help. [. ...] This is an absolutely excruciating experience for brilliant, intellectual people. They tend to freeze when they hear the whistle and to try to analyze what they were doing. That they don't know, and that it doesn't matter that they don't know, is a shocker" (K. Pryor 1984, 68–69).

that made it possible: neo-liberalism has fooled us into believing that choosing from among several options constitutes thinking and is thus a guarantee of freedom. The (human or canine) subject experiences a sense of satisfaction and accomplishment as a result of believing that he had arrived at the solution himself.

According to Fisher's *The Thinking Dog*, allowing the dog to "think," to fully comprehend the complexity of the task, is the best way of ensuring his performance. "Thinking" here is seen as resulting in the choice of the most beneficial option. Gail Fisher wrote: "A 'trained' dog will obey, but a thinking dog goes far beyond obeying commands. [. ...] It's about when a dog is faced with options, he considers them and chooses the right one – the good behavior. A thinking dog chooses to behave the way you want him to behave not because you gave him a command" (Fisher ix). Thinking is choosing from among several possible options. The rhetoric of choice, so typical of neo-liberal governmentality – as *Žižek* puts is "at the very nerve center of the liberal ideology" (*Žižek* 2001, 114) – seeps its way into dog training offering empowerment to the dog. Choice becomes so firmly associated with reinforcement-based training precisely because it is the paradigm of choice, so to speak, for the neoliberal subject who sees him- or herself as shaping his life project by choosing freely from among many different options. The old paradigm, the paradigm of obedience, insisted firmly that the dog could not be given a choice. That said, early on in the history of reinforcement-based training some trainers were aware of the inconsistencies of associating choice only with the positive training camp. In a very perceptive meta-comment on the rhetoric of choice as an artificially crafted division line between the two training camps Jean Donaldson wrote: "I once heard a competitive trainer characterize the use of force as a justifiable means to reduce the dog's perception of choice. But this doesn't even work in theory: the dog always has a choice, even if it's between two pretty scary options, one of them cooked up by the trainer" (Donaldson 1996, 37).

It has been Skinner's work that has informed contemporary animal training but the similarity between the intellectual legacies of these two thinkers, Skinner and Foucault, makes it particularly appealing to use Foucault to discuss the changes in contemporary dog training in order to situate them within the broader framework that Foucault offers. Conversely, reading Skinner through Foucault makes it possible to see him not as a slightly loony visionary with a very reductionist interpretation of human life, which is how he is generally treated today, but as a diagnostician of the change that has shaped the modern neoliberal society: the move away from the reliance on a primarily disciplinary – and, at the same time, punitive – model of social control and towards one based on influencing the will (rather than just the body) of the subject.

Deleuze wrote of the move from the model of the factory to the model of the corporation (1992). Foucault wrote of the development of modalities of power in terms of efficiency but did see a place for affect (and affection) within changes in power regimes. Conversely, Skinner advocates for positive reinforcement not solely on the basis of its humaneness, but proposes it as a more effective way of shaping the behavior of humans and nonhumans alike. This undiscussed similarity between Skinner and Foucault – bearing in mind that Foucault reads Skinner as a neo-liberal theorist – probably underruns some of the discomfort that certain Foucauldian critics sense in the relationship of Foucault to neo-liberalism; by discomfort, I mean that they view Foucault as not critical enough of neo-liberalism (Gordon 1991; Zamora and Behrent 2016).

2 Training as a Practice of Freedom

Foucault's work has, of course, been instrumental in the dismantling of the notion of the autonomous subject within the debate that practically defined the humanities in the last decades of the twentieth century. He had spoken of the objective of his work being "to create a history of the different modes by which, in our culture, human beings are made into subjects" and insisted that it was the subject who was at the center of his life's project (Foucault 1982, 778), even though there is a certain shift of focus from how the subject is shaped by technologies of power and domination in Foucault's earlier work to self-shaping of the subject within the constraints of existing power regimes in Foucault's later work. Foucault's acknowledgment of how the subject is shaped by power is never tantamount to an outright rejection of the notion of freedom. For Foucault, freedom is not an abstract category, it is a material practice that inevitably takes place within a certain power apparatus (Fornet-Betancourt et al. 1987). In his insistence on the material character of freedom and in defining it through behavior, Foucault again crosses paths with B.F. Skinner.

Skinner is known for his focus on the tangible, the observable, the material: in other words, on behavior as opposed to thought. He used the term "mentalism" as the ultimate insult. For Skinner, every concept, no matter how seemingly abstract, can be defined in terms of contingencies of reinforcement. In Skinner's understanding of the world, the subject's behavior is shaped by the environment, which reinforces certain behaviors and punishes others in ways that the subject – whom Skinner refers to as the organism – is usually not aware of. Freedom, in the understanding of being free to do whatever one wants, is thus a fallacy because one can never do what one truly wants (Skinner 1971, 31–32). The feeling of freedom is an illusion, which can be induced in individuals through contingencies of reinforcement. Similarly to Foucault, Skinner thus recognizes the reductionist and incomplete character of the enlightenment reading of man as the autonomous individual even if, paradoxically, it is his science that is seen as the "Enlightenment" of dog training. The kind of freedom that interests Skinner and that he does not read as an illusion but as, historically speaking, the driving force behind social and political change is a behavior: the attempt to free oneself from aversive stimuli. However, this definition brings to the surface a potential problem: if the conditions of control are not aversive, the individual will not desire to free himself from them even if they are not in his best interest. Skinner argues: "The feeling of freedom becomes an unreliable guide to action as soon as would-be controllers turn to non-aversive measures" (32). Again, the similarity between Skinner's aversive and non-aversive measures and the situation of the subject under Foucauldian disciplinary power versus neo-liberal governmentality is strikingly similar. The correlation of aversive power and disciplinary power is not exact, but the similarities are glaring: the primary techniques of discipline focus on the body, instilling habits that affect the mind via the body. This is a somewhat passive process in which the subject is shaped. Because post-disciplinary techniques (using the term very broadly) focus on affecting the subject's desires, the process of shaping the behavior desired by the controller becomes a much more active one for the subject. In other words, once we move away from discipline, resistance to power becomes much more difficult.

Unlike Foucault, however, Skinner does not dwell on the dangers of non-aversive social control and he channels his intellectual effort into promoting non-aversive measures of behavior modification. The potential benefits of shaping human behavior using positive reinforcement drive his work over a period of several decades: from *Walden Two* (1948) to *Beyond Freedom and Dignity* (1971). It is this effort that generates the greatest resistance to Skinner's ideas. The accusations are obvious: if the controllers do not have the best interests of the controlled in mind (and sometimes, even if they do), positive reinforcement can become a powerful tool of "mind control" (Hocutt 2013). If positive reinforcement can be used to train pigeons to carry out suicide missions, the possibilities for abusing the power of positive reinforcement as applied to humans are indeed endless. Skinner, however, as always down to earth, responds that these contingencies of behavior are universal scientific principles that should not be hidden from the public eye for fear of their misuse. In fact, the best possible defense against the misuse of positive reinforcement as a tool of social control is an understanding of how positive reinforcement works, which provides both the means of spotting and identifying reinforcement protocols at work and the possibility of self-management. As time passed and radical behaviorism gave way to cognitive psychology, it was behavioral therapy with its emphasis on self-management that provided a path for retaining Skinner's vision. As Nikolas Rose writes in *Governing the Soul*, the name given to this new field of application is "self-control": "the systematic management of environmental circumstances and response-contingent reinforcement by the clients themselves" (Rose 1999b, 241). And again, the idea of self-management brings Skinner in conversation with Foucault.

The last chapter of Skinner's biography by Daniel W. Bjork is titled "Master of Self-Management" and describes how in the later decades of his life B.F. Skinner used the principles of operant conditioning to arrange his environment in a way that would heighten his own intellectual productivity. He kept to a tight schedule, isolated himself from external interference, yet allowed appropriate social interactions that could prove reinforcing. As his health began failing, he increasingly relied on technology to allow him to live what he called a productive life. He even designed a "thinking aid" in the form of a filing system that helped him organize the thoughts he found increasingly more difficult to order (Bjork 1993, 223). As Bjork writes of Skinner: "He viewed himself as the organism, shaped by the effects of environmental contingencies he knew best" (224). In his last years, Skinner focused on self-management also in his writings and included a sub-chapter titled "Managing Oneself" in *About Behaviorism* (1974), the follow-up book to *Beyond Freedom and Dignity*. He had previously included a chapter titled "Self-Control" in *Science and Human Behavior* (1953). Skinner even wrote a self-help book titled *Enjoy Old Age: A Program of Self-Management* (1983) and lectured extensively on self-management in both academic and non-academic contexts. In other words, as Skinner ages, his interest moves from examining the role of coercion and positive-reinforcement in behavior modification to technologies of self-formation: a shift that eerily mirrors Foucault's change of interests in the last decades of his life. In fact, from a Foucauldian perspective, Skinner seems to have been engaging in the practice of technologies of the self, a concept that Foucault analyzed in his later years, though in greatest detail in the second and third volume of *History of Sexuality* (originally published in French in 1984), which were written after his *Birth of Biopolitics* lectures (delivered in 1978–79), where he exhibited an awareness of Skinner's work.

Skinner's self-management, based on an internalization of the principles of operant conditioning seems quite similar, both in terms of the technologies through which it is carried out and the goals of self-management, to Foucault's technologies of the self. In terms of goals, both Foucault and Skinner see self-management as a practice of freedom. Of course, not of the (impossible) absolute freedom of the autonomous subject but of the kind of freedom available to the subject of power within the constraints of the power apparatus he is functioning in. Skinner, of course, used a different vocabulary and spoke of an organism influenced by the environment but the gist of his idea is quite similar to Foucault's. In his later years, Skinner also acknowledged the role of the body, which had to be carefully managed in order to extend its productivity and thus, his techniques of self-management include careful regulation of the body's activities (making sure that one gets enough sleep, proper food, light and exercise). Engaging in self-management allowed Skinner to experience freedom in a way in which he believed would not have been possible had he allowed his mind and his body to deteriorate. In a sense, then, self-management was also an aesthetic practice, just like it had been in Foucault's reading of the Greeks' vision of an "aesthetics of existence"; a strategy for leading a beautiful life. Skinner's vision has been called that of "the good life" in the title of a 1979 biographical movie about him and in recognition of the similarities in the thought of Skinner and Epicurus (Neuringer and Englert 2017). It was a life free of temper tantrums, useless quarrels but – above all – an intellectually productive life.

In terms of technologies, the similarities are even more striking and center around controlling the environment in a way that makes it possible to increase one's possibilities of withstanding temptation. The Greek's notion of *epimeleia heatou*, a habit of "taking care of oneself," the practice of correct responses in a controlled environment, with the goal of internalizing the response to the extent that it becomes automatic. In a 1984 interview on the topic of the care of the self, Foucault thus comments on Plutarch: "You have there the idea of a *logos* who would operate in some way without your doing anything. You will have become the *logos* or the *logos* will have become you" (Fornet-Betancourt et al. 1987, 117). Ethical work, for Foucault, consists of the practice of self-forming activities and finally results in the self-shaping of oneself into an ethical agent.

3 Doggie Zen: Dog Training and Technologies of the Self

The application of behaviorist techniques in dog training also has as its goal the formation of a self-regulating canine: one who has internalized the "good behavior" to the extent that it becomes the default response. This is reflected in the exercise that seems the most representative of reinforcement-based training, in the same way that obedience exercises such as heeling, staying and coming when called were associated with the previous regime of power. Known under the various guises of doggie zen, self-control, impulse control and "It's Your Choice," for reinforcement-based trainers this exercise is not just a flashy trick but a foundational life skill for dogs. Even though there exist differing instructions for teaching this behavior, all trainers are in agreement about its goal, which is to inhibit the dog's instinctive and impulsive reaction in the face of a tempting stimulus. While B.F Skinner can be called the precursor of the technique (not to forget S.T. Hammond's version of the exercise from the 1880s), the earliest mention of impulse control exercises specifically in reinforcement-based dog training literature can be traced to the web-based writings of Canadian trainer Sue Ailsby, who, in the mid-1990s, called the exercise "Doggie Zen." In her books, blog posts and online training materials Susan Garrett, a top-class Canadian trainer whose popularity borders on stardom in the dog training world, calls her version of this exercise "It's Your Choice," a name which emphasizes the dog's active role in the process, and considers it one of the "foundational games" for training regimens geared towards multiple canine sports (Garrett 2005, 71–74). The exercise consists of showing the dog a treat, held in the palm of the trainer's hand, and refusing to give the treat to the dog as long as the dog retains interest in the treat by nudging the hand, pawing, etc. Once the dog understands that attempts to access the treat do not yield the desired result and stops them, the hand is opened, and the treat released to the dog. As Sue Ailsby writes:

> Zen, as we use the word in dog training, is so important as to be virtually the foundation of civilization. It means "self-control." An untrained dog is a dog with no self-control. She wants food; she eats food, whether that food is on the floor, on the counter, or in a toddler's hand. (….) A trained dog understands that the way to get what she wants is to control herself, and a trained handler knows that true control of an animal must come from the animal herself, not from the handler. (….) The great thing about doggie Zen is that once the dog has learned enough of it, she starts to apply the principles of self-control to her entire life. She practices "leash Zen" by keeping the leash loose, "floor Zen" by ignoring bait on the floor in the show ring, and "table Zen" by sitting and "wishing" food off the table instead of jumping up to help herself. (Ailsby 2011)

I emphasize Ailsby's assertion that the purpose of the self-control game she refers to as Zen is not just the successful performance of this exercise but that it is meant to be generalized by the dog in other contexts. After the training process is completed, Ailsby expects the dog to control his impulsive reactions when faced with various tempting stimuli (Figure 4). Ailsby calls Zen, or impulse control, "the foundation of civilization" and this wording is poignant in light of the topic she is discussing: human-canine relations. Impulse control thus becomes an element of the social contract between human and canine. Furthermore,

FIGURE 4. In positive training, exercises in which the dog is taught to resist the urge eat a tempting piece of food - known as self-control, impulse control or doggie zen – are taught without resorting to physical punishment. Still, self-control is considered a value.
STILL IMAGE FROM VIDEO INSTALLATION "IN MEMORY OF ALL THOSE WHO WORK WITHOUT GETTING A REWARD" (2007) BY LISA STROMBECK.

Ailsby emphasizes that true control is self-control, it must result from the animal; it is the practice of impulse control that results in the dog developing a desire to act in a controlled manner.

Targeting, or touching a non-baited object – e.g. the end of a long stick (Spector), a plexiglass square (Garrett) or just the handler's hand (Miller, Garrett) – is also a popular exercise based on the same premise as doggie zen: the dog needs to learn that the only way to obtain the treat is to stop focusing on the food and do something else: to nose touch an object. Both of these exercises are seen as foundational because they introduce the dog to the concept of delayed reinforcement – after all, the very point of introducing a marker, like the clicker, is to be able to postpone the delivery of reinforcement – but also because they both require the dog to give up the most obvious path to reinforcement. This is also why the majority of reinforcement-based trainers value the technique of shaping behavior over luring, even though both these techniques do not rely on positive punishment. In luring, the dog follows the treat in order to obtain the position desired by the trainer. In shaping, the dog does not follow a treat; instead, his successive approximations of the desired behavior are marked and reinforced. Shaping is seen as the true technology of positive training: offering more freedom and flexibility in teaching behaviors (not all behaviors can be lured) but also as inculcating in the dog good "thinking habits": the dog learns that the availability of the reinforcer is contingent upon his behavior, not on the physical presence of the food or toy.

While these exercises are not consciously involved in a discussion of animality, it is striking that the kind of "good behavior" that the dog is expected to choose on his own, is marked as defying "animality." Foucault reads Epictetus' *Discourses* as locating the difference between man and beast in the belief that man is the being who was "destined to care for himself" (Foucault 1985, 47). Animality is the instinctive, impulsive reaction: the animal does not wait but follows his impulses, unable to control them: the animal satiates his appetite immediately upon seeing food. He cannot help it. Defining animality in terms of lack of self-control has a long history in Western thought and is by no means limited to the Greeks. In *The Autonomous Animal: Self Governance and the Modern Subject* (2011) Claire Rasmussen, through a detailed reading of the notion of animality in both philosophy and political theory, argues that within much modern political thought moral agency has been synonymous with the capacity for autonomy; the ability to act in a rational manner, as a result of reason and not impulse. Chapter 1 argued that already in the nineteenth century training seemed to provide an opportunity for partial transcendence of animality for dogs. Per Foucault's reading of Plutarch and the Stoics, an important element of the general doctrine of self-care consists of changing the spontaneous/animalistic reaction through the inculcation of "truths" and "fundamental principles." As a result, when one's appetites awaken, the instinctive reaction will no longer be the animalistic one, that is immediate satisfaction of one's desires, but a different response based on internalized principles. For the Stoics, the ethical response was neither a renunciation of immediate gratification with the hope of delayed gratification in the afterlife – as Foucault explains, it is Christianity with its concept of salvation which shifts the focus of self-care towards the afterlife – nor necessarily a sacrifice of individual pleasure for the greater benefit of the society (Fornet-Betancourt et al., 116). It could be a temporal delay of immediate pleasure practiced for the sake of building one's character. This work to be performed by the subject who desires to be an ethical agent is that of self-mastery. This is obtained through asceticism, which, however, was not understood as ultimate renunciation of all bodily pleasure (Fornet-Betancourt et al. 1987, 113; Foucault 1986, 68). Ascetic practices are, indeed, focused around self-restraint and self-discipline but they aim neither at eliminating one's desires for physical pleasures nor the pleasures themselves. Self-mastery is achieved through careful and conscious use of the pleasures. The exercises themselves often resemble that of "doggie zen." In the interview "On the Genealogy of Ethics" Foucault argues that "for Epictetus, or for Seneca, for instance, in order to really be master of yourself, you don't have to know what your role in society or in your home is, but you do have to do some exercises like depriving yourself of eating for two or three days, in order to be sure that you can control yourself" (Foucault 1984b, 358)

In the Greek version – and somewhat by default in Foucault's work – self-mastery is available only to human beings. Skinner's universalism – his extension of the principles of learning theory onto the animal world – truly changes the game, and it is this change that thrilled dog trainers. If operant conditioning works on all organisms, then dogs become capable of self-control as well, and this is what the exercises described above provide instructions for. These are ultimately games of character-building for the dog, just like Foucauldian technologies of the self were for people. While the attainment of self-control, control over one's animalistic impulses, was seen as the goal of dog training already in the nineteenth century, the mechanisms for achieving this state were interpreted differently then. Chapter 2 analyzed how the bird dog's self-control was linked to his good breeding: "the genteel" dog was not impulsive because of his superior heritage, which had been influenced by humans through the choice of appropriate mates. Konrad Most redefined the role of instinct in a Nietzschean fashion and based his training system on suppressing some canine instincts and the development of others; all under hierarchically-understood human control. Even if self-control was achieved, it was seen as passive acquiescence to the intellectually and physically superior human. Something different happens with the adoption of Skinner's operant conditioning in dog training in the 1990s. The concept of self-control replaces obedience as the conceptual underpinning of training. The possibility of attaining self-control, which is still equated with rationality and thinking and defined as the goal of training, becomes the potential province of all dogs if only they are subjected to appropriate training procedures. Nikolas Rose writes of the application of Skinnerian principles: "self-control might sound like the reactivation of old moral nostrums, and to some extent it is. But the difference is that it is not merely a matter of 'will-power' but of the systematic management of one's natural and social environment ('stimulus situations') in order to transform the likelihood of pleasurable concomitants of particular types of conduct ('response contingencies')" (Rose 1999b, 241). The incorporation of the notion of self-control into dog training reflects the desire to appreciate and acknowledge canine subjectivity. If we see dogs as capable of rational action – deferring pleasure and acting against their "animalistic" impulses – then our opinion of their potential as subjects of self-governmentality certainly increases. However, this approach still relies on a very obvious division into rational and non-rational actors. It is rationality, originally the province of humans, which is extended onto dogs (albeit only to a limited extent). It is a notion which is still anthropocentric – and humanist – at heart: it is the dog's potential of adjusting to the human's definition of rationality that makes him worthy of appreciation.

4 From Discipline to Control

For a technology of love, reinforcement-based training put surprisingly little emphasis on touch. In the slightly parallel world of horse training, the turn to "soft" technologies of power has engendered new ways of experiencing the human-horse interaction, largely described through an intensification of the bodily connection between horse and rider. The founding figures of these movements, Monty Roberts and Pat Parelli, eagerly explored the various forms of touch between human and horse. Notions such as co-embodiment and mutual corporeality also seem to be productive for theoretical explorations of the practices of natural horsemanship and horse whispering: soft power is about feeling the body of the other and responding to the presence of that body. Lynda Birke and Keri Brandt call for a "mutual corporeality of person and horse to explore how our engagement with the animal produces a co-embodiment" (Birke and Brandt 2009, 196). It is not surprising that affect theory has proved useful for conceptualizing practices of embodiment in the world of horses. Elspeth Probyn, playing with Deleuze and Guattari's "becoming animal" writes of "becoming horse," the affective connection between the woman and the horse as modulated via bodily experience (Probyn 1996, 56–60). Attempts at redefining the human-horse relationship in natural horsemanship entailed experiencing new forms of tactile contact. In fact, bodily encounters devoid of bits, reins and saddles became explicitly more corporeal as the contact between the body of the horse and the body of the rider was less mediated via the various props for controlling the animal's behavior. Not so with dogs. Even if the training apparatus, consisting of choke chains, prong collars and the leash is abandoned, it is not immediately replaced with a new and more immediate form of touch. On the contrary, reinforcement-based training develops as a hands-off approach. "Don't touch the dog" could most certainly be a motto for clicker training.

In dog training the turn towards postdisciplinary control is a turn away from human-canine bodily interaction and towards what is perceived as an emotional connection. This may seem surprising if we think about training as a technology of love: recognition of the strong emotional attachment of the human to the canine and the concurrent strengthening of the dog's attachment to the human. One would think that the love professed by the human should logically result in a desire to touch, to pet, to stroke, to have the dog's body respond in reaction to the human's loving touch and to incorporate these responses into the training regimen. The reasons why touch did not become an element of reinforcement-based training are twofold and have to do with the legacy of coercive touch as a constitutive element of the disciplinary regime and, simultaneously, with the nature of postdisciplinary control as targeting the subject's affects rather than the body. As regards the first reason, refraining from touching the dog in the process of training is seen as a break with the traditional training techniques that targeted the dog's body. The perception of touch as an element of traditional training was evident in Spector's exclamation about throwing away the leash and the collar. This simple gesture had huge liberating potential specifically because it broke with the long history of touch combined with force. The connection between interspecies touch and coercion was much stronger in the disciplinary regime as it pertained to canines than as it did to humans. For obvious reasons, dog training could not rely on some of the techniques of human disciplinary institutions and the physical positioning of the animal's body in the desired position was much more important than it had been in the context of disciplinary practices for humans. As a result, not only did all interspecies touch begin to be perceived as coercive by the change-oriented trainers but, by extension, Skinner's equation of coercion with punishment made a lot more sense for the animal trainers than it did in its application to human relations. It was an unquestioned assumption of clicker training in the 1990s that if one removes aversive stimuli (Skinner's positive punishment) from the repertoire of response contingencies, we can no longer speak of coercion. Corporal punishment in schools is an obvious equivalent of corrections in dog training, yet one would be hard pressed to find someone who believes that removing corporal punishment has done away with all coercion in schools.

It certainly did not help that dog training had a legacy of "bad touch"; touch that was deceptive in its apparent friendliness and that easily turned into physical coercion: the handler's encouraging chatter coupled with a cunning "kick correction" that puts the dog in position; Saunders' trademark coupling of jerking and praising. Wariness of touch is also a response to training techniques that develop in the 1970s and that proclaim to focus on building the human-animal bond – thus showing a modernizing focus, less emphasis on utility, more emphasis on training as a relationship-building program – but that turn to physically harsh corrections, grounding them as "only natural." These techniques are discussed in more detail in chapter 6 but a manual that is exemplary of this mode of thinking is *How to Be Your Dog's Best Friend* (1978) by The Monks of New Skete: a book that has interestingly become synonymous with traditional training, even though it is actually one of the first manuals that focus to a greater extent on the human-animal bond than on providing detailed "how-to" instructions. The Monks, however, advocate the technique of punishing the dog by rolling him over onto his back, exposing the dog's belly and pinning the animal to the ground until he stops fighting the procedure. This maneuver was referred to as the "alpha roll" and was promoted as a technique that imitates the mother dog's correction of an unruly puppy. In the illustration accompanying the description of this technique in the book, the man pushing the dog into the ground has a wistful, almost caring expression on his face, as if he was connecting with the spirit of the dog's wolf ancestors (1978, 47). For the clicker trainers, Skinner's "hands off" approach – which for Skinner was simply an effective technique – had the additional benefit of disengaging from the legacy of coercive touch.

While many of the early clicker training advocates compensate for the lack of touch in training through touching techniques carried out outside of the context of training, such as special massage techniques (Fisher, Spector), for other positive trainers the absence of direct physical contact also becomes a sign of the respectful recognition of the dog's otherness. Skinner's antimentalist stance transpired into caution with regards to the imposition of humans' mental states onto dogs: including the assumption that touching is necessarily pleasurable for the dog. This wariness is written into protocols of reinforcement that never treat petting – the classic reward in traditional training – as a reinforcer and that make sure to classically condition a dog's acceptance of human touch (Spector 5, Fisher 70). Turid Rugaas, a Norwegian trainer famous for her analysis of canine body language, is even known for her warning that one should never hug one's dog (Homans 2012, 44–45); a warning reiterated by behaviorist and trainer Patricia McConnell (2002, 16–19) and one that recently made headlines in the popular media after Stanley Coren's study on signs of stress in pictures of dogs being hugged by humans (Coren 2016b).[2]

The second reason for the turn away from touching the dog's body can be explained with the embedding of dog training into the more general shift from disciplinary to postdisciplinary power. These reasons are by no means disconnected but the structural reason deserves an even closer look. Foucault's turn away from discipline and toward biopower has been picked up and developed by more recent theorists who, while not negating the importance of biopower in the contemporary world, propose a different terminology that refocuses the shift in modalities of power at the turn of the twentieth and twenty-first centuries to conceptualize how the crisis of discipline is not equivalent to a loosening of social control over the subject of power. In a short article published in 1992 Deleuze outlines the concept of "societies of control" – as he claims, extending the thought process initiated by Foucault – and the term is later picked up by Nikolas Rose in *Powers of Freedom* (1999a) and by Hardt and Negri in *Empire* (2001). I find the notion of a "control society" more productive in relation to a discussion of both the methods and the functions of dog training, as it perfectly captures the gist of the changes: specifically, the shift from discipline's focus on affecting the individual body to the postdisciplinary techniques of modulating affects. The techniques of discipline mold and shape the body into desired poses and positions. The techniques of control focus on creating motivation for the subject to behave in particular ways and not others. As feminist scholar Patricia Clough writes: "[t]he target of control is not subjects whose behavior expresses internalized social norms; rather,

2 It should be noted that Coren's article was almost immediately rebuked by Marc Bekoff who criticized Coren's overarching conclusion. Turid Rugaas is also a somewhat controversial figure in the world of dog training: an anti-trainer among trainers. She never adopted clicker training but also vowed to never use punishment in her interactions with animals. She is known for her work on canine body language, signs of discomfort and stress (calming signals) and, more recently, on stress levels. She is controversial for her opposition to all activities that increase a dog's cortisone levels: including ball throwing, agility training, etc. However, the fact that reinforcement-based training in the 1990s saw the work of Rugaas as an extension of the work of clicker training is in itself significant. Rugaas was invited to seminars organized by clicker training facilities and her work on calming signals was routinely referenced on internet mailing lists. These early clicker trainers truly wanted the best for their dogs and saw Rugaas's wellness approach as complementary to the more technical approach of positive training.

control aims at a never-ending modulation of moods, capacities, affects, potentialities" (Clough 2003, 360). Skinner's behaviorism, particularly as applied in the training of pet dogs, is the exemplary technology of control: not only does it affect the subject's motivation for performing certain behaviors, it can actually be carried out without physical interaction with the subject. Finally, the loosening of disciplinary control of the body does not mean a loosening of control in general and, most certainly, does not entail a loss of the body's productivity. As stated in the previous chapter, all of the early clicker trainers emphasize that their way of training is even more effective than traditional training techniques.

The other features of a society of control include a crisis of the classic disciplinary institutions (Deleuze 1992, 3–4), a changing understanding of space and loss of significance of distinct spatial enclosures (4), the replacement of the school by perpetual training (5). The society of control, it seems, would not be possible without developments of technology: in the 1990s Deleuze speaks of the home office, but it is clear that the advent of the internet with the endless possibilities of tracking the behavior of internet users is absolutely key. The changes within dog training can be read in the context of Deleuze's society of control. The disciplinary institution for canines, the dog obedience school, clearly goes into crisis mode in the 1990s. Not only do large group obedience classes go out of fashion along with the advent of clicker training, to be replaced by individually-tailored behavior modifications programs implemented in the dog's home surroundings, the overall reputation of the obedience school certainly suffers as it loses its strong tie with national ideological apparatuses of governmentality. In the twenty-first century, one would be hard pressed to find a dog obedience class that meets in a church or in a school. The words "Obedience School Dropout" can now be found on the bumper stickers of cars owned by many dog lovers as a humorous sign of disidentification with this disciplinary institution. According to research published by Stanley Coren in 1999, only 24% of dogs in the United States attended formal obedience classes (Coren 1999, 199–200). Reflecting on why this figure is so low, Bennett and Rohlf suggested in 2007 that "perhaps many owners simply train their dogs at home, accessing training knowledge via books, videos or informal conversations with friends and family members" (Bennet and Rohlf 2007, 66). Hiring a trainer who individually attends to the dog in a home setting should also be added to the list. In a UK-based study from 2008, 88% of dog owners responded that their dogs have had "some training" (Blackwell et al. 2008). Clearly, it is not training that is in crisis but the concept of the dog obedience school with its Koehler-style huge classes. On the contrary, the training of pet dogs is more popular than ever but rather than being a procedure carried out within the confines of the school, during an eight-week-long course, it becomes embedded in daily human-canine interactions. The focus on teaching behaviors geared towards obedience competitions also wanes. Instead of heeling, coming to front and finish and retrieving over a hurdle, pet dogs begin to be taught to walk on a loose leash (the position being much less precise than in the obedience regulations), politely greet guests (not jump up) and maintain their cool in the presence of squirrels, cats and cars (usually through self-control).

The market for pet dog training has been taken over from the non-profit clubs that dominated since the 1940s to the 1990s by private businesses, which cater to the needs of the individual human-canine pair. Even if a dog attends a training class, the majority of the work is not taught in class but outside of it. While already Saunders spoke of homework as a necessity for obedience school students, homework was understood as the repetition of exercises first taught in the group context; the legwork necessary for instilling good habits. As training classes become redefined, they begin to serve more as a laying out of the theory of training, which needs to be put into practice outside of the context of the class. Disciplinary institutions operated under the assumption that once training is completed, the subject's behavior has been altered in a permanent way. Attending obedience school results in a trained dog: the product of disciplinary training. As discipline gives way to control, and at the same time as training becomes redefined as relationship-building, expectations change. Behaviors such as sit and down are no longer considered products of training but steps in the program of human-canine relationship-building. Skinner's analysis of behavior and particularly his technique of breaking behaviors down into discrete components, leads to an understanding of the multiple criteria that go into the performance of a particular behavior. Students begin to be taught that the generalization of a correct response to a cue is always a life-long learning process. Pryor warns that a new environment always constitutes a new criterion (K. Pryor 1984, 84–86) and thus the process of proofing the performance of the desired behavior through the raising of criteria is potentially endless. Very literally, training becomes perpetual. Pat Miller warns that "[y]ou may need several years of committed training in order for your dog to be 90% reliable around very strong distractions" (Miller 2001, 73), which is a far cry for Saunders's eight-week-long course in general obedience that promised a dog who would always respond to the "Come!" command. However, this really does not matter because the goal of pet dog training ceases to be perceived as a reliable response to a command but as the deepening of the human-canine bond. The crisis of the obedience school is linked to a redefinition of training. If training begins to be, as Spector argues, about creating "a feeling of community and 'family'" (Spector 1999, 2), it cannot be accomplished in the context of a large group class that meets once a week.[3]

Theorists of the society of control insist that technology has been instrumental in the move away from disciplinary power: Hardt and Negri are particularly explicit about this throughout *Empire*. Technology also plays a role in the changes in dog training. Firstly, reinforcement-based training is perceived and marketed as a technology; the first and only science of learning, presented in a framework of civilizational and technological progress. The idea of civilizational progress is based on a construction of "the dark ages" set against the current age of science; as Foucault reminds us in "What is Enlightenment?," enlightenment is defined by a modification of the preexisting relation linking will, authority and the use of reason (Foucault 1984d, 32–50). A binary of "before" and "after" is thus constructed with everything that came "before" necessarily marked as barbaric.[4] The visual representation of technological progress is a clicker: a conditioned reinforcer whose value over a verbal marker is seen in its neutrality. Spector posits that the use of the clicker is "integral to the training program" because "its sound is distinct; its sound is uniquely paired with the presentation of the primary reinforcer; it is immediate; it is audible over a distance; and it is neutral, that is, it is unaffected by emotion or physical factors" (Spector 1999, 8). The objective neutrality of the scientist (or the clicker trainer) is set against the legacy of subjective, "non-scientific" solutions that inevitably turned toward punishment even if they professed gentleness.

---

3 It must be noted that there exists an obvious lack of cohesion between the emphasis on the effectiveness of clicker training as a scientific technology of training and the approach that results do not really matter because it is the bonding process that counts. This disjunction is often discernible in the same book. Both Spector and Miller write of clicker training's superior effectiveness only to emphasize that the relationship-building aspects are really more important. There also seems to be a pattern of trainers becoming less competitive after switching to clicker training. Miller writes: "Josie never got her utility title. Somehow it never again seemed that important." (9).

4 As a result, if there is anything that clicker training is guilty of it is the wholesale equation of all training that came before the *conscious* application of Skinner's operant conditioning to the training of dogs as barbaric. The amazing paradox is that it is not interested in historicizing the concept of positive training because it sees it as a technological innovation. The actual existence of trainers who clearly understood operant conditioning before Skinner and who chose to rely on positive reinforcement (S.T. Hammond is an obvious example here but not the only one; Jay Sisler, who trained dogs who performed during rodeos in the 1950s is another) is pushed into oblivion. This rhetoric of technological progress also explains clicker trainers' vehement disagreements with "traditional" trainers. The intensity of "the clicker wars" that took place mostly on internet discussion lists in the late 1990s resulted from the disdain that the proponents of operant conditioning had for all other trainers.

Through its association with objective, scientific knowledge, the clicker is seen as a guarantee of the gentleness that had been sought by trainers for decades. Indeed, it constitutes such a guarantee once (and only if) gentleness becomes "scientifically" defined via Skinner's quadrant as absence of aversive (painful) consequences. Finally, the development of new channels for communicating among humans, specifically the popularization of the Internet, was absolutely key for the outburst of interest in clicker training in the second half of the 1990s. Even though published books on the topic were still lacking, internet discussion lists became the venue where positive methods were discussed, popularized and the site of endless wars between clicker trainers and proponents of "traditional" training methods.[5] In her acceptance speech for the 1997 Annual Award for Excellence presented by the Association for Behavior Analysis, Karen Pryor said:

> Five years ago, I could not have named a single clicker trainer in the dog-owning community. Not one. Now, I am personally aware of over ten thousand adherents to clicker training. Our company, Sunshine Books, has a website offering a catalog of clicker training books, videos, and gear, as well as news, downloadable teaching materials, articles, and so on. Last month alone (April, 1997), our website had 20,000 hits. (K. Pryor 1997)

Clicker training was both marketed as an innovative technology and spread along with the development of the technology of cyberspace. In conclusion, behaviorism – and particularly the technology of clicker training that sprung off from Skinner's ideas – was surprisingly compatible with the broader turn from discipline to control that was taking place on the level of power modalities.

5 Accounting for Affect/Accounting for Gender

The Deleuzian analysis of the "society of control" does reveal the naiveté of the trainers who viewed behaviorism as a methodology that offered liberation to the dog, but this reading does not, in any way, discredit the significance of this

5 The author remembers the intensity of these discussions in the late 1990s from personal experience. The most significant English-language internet discussion groups included: Click-L (the first such list, operated in the 1990s by Kathleen Weaver), Clicker Solutions (operated by Melissa Alexander), Click for Success (operated by Lana Mitchell), ClickerTrain (operated by Corally Burmaster), ClickTrain (operated by Shirley Chong). Most of these communities lost their momentum in the early 2000s and none remain active today.

change and the strength of the trainers' convictions: it simply does not account for them. The analysis of the shift from discipline to control provides an understanding of how the dog is affected as a subject in the process of training and of how training itself fits into the broader framework of the shift away from discipline. The one issue that fails to be accounted for in this reading of how dog training changes as it moves towards postdiscipline is the trainer: in particular, the character of the attachment that the trainer experiences when training the dog; the modulation of affect in the trainer. In various theorizations of postdisciplinary power, the analyses pertain to relationships between an individual subject of power and an impersonal entity. Deleuze speaks of the corporation as the model for the society of control, and we can imagine the relationship as that between an actual employee and a faceless yet skillful corporate power. Yet if we think of operant conditioning as akin to postdisciplinary power – which I have argued is a valid approach even if the analogy is not exact – the variables change because the kind of power Skinner writes about can be a relationship between two people (though it does not have to be; it can also be a principle governing an institution). Both Skinner, in *About Behaviorism* and *Beyond Freedom and Dignity*, and Karen Pryor, in *Don't Shoot the Dog*, posit that the power of behaviorism is a power to change relationships (as it changes behavior). Many of the examples Pryor provides in her 1984 book are examples of interpersonal relationships in which the application of positive reinforcement to modify behavior results in increasing the affection of the "trainer" (quotes are used here because Pryor writes about, for example, shaping the behavior of one's spouse) for the subject of training. Pryor gives specific examples for dealing with a grumpy spouse and a tantrum-throwing child. She argues that "positive reinforcement is good for human relationships. It is the basis of the art of giving presents" (1984, 23). Her solutions result not only in changing the behavior of the subject of training but also in increasing the affection of the trainer for the trainee. It is, after all, easier to love a husband who is not grumpy and a child who does not throw tantrums. It is also easier to love a dog whose behavior is predictable, controllable and subject to modification.

However, in the case of dog training, the affection is not only the end point of training but also the starting point: "feeling power" is what drives this change, as is clear from the analysis of the crossover memoirs. In turn, the new methods are seen as deepening the trainer's love and creating a reciprocal affection in the dog: it is specifically this reciprocity that was seen as lacking in the old methods. It is this affection that made it possible for trainers to recognize and avoid many of the pitfalls of other apparatuses of control, such as the corporation, and that eventually led toward further transformation of pet dog training away from radical behaviorism. The goal of control in a corporate setting is increased productivity of the employee, which ultimately results in higher profits for the employer. As such, control can explicitly proclaim care and love for the employee, while simultaneously turning into a highly exploitative tool. Control, after all, is a modality strongly associated with a late capitalist society, which is adept at maximizing the profits of those who are already economically privileged. This does not seem to be the case in the kind of training relationship I am describing here. The bidirectionality or reciprocity of the affective connection, its honesty and lack of a profit motive, result in a situation that creates a set of checks preventing the exploitation of the animal. It is very rare, practically unheard of, for purely positive trainers – even those who are involved in competitive training – to approach training cynically: to increase the animal's affection with the goal of abusing the dog's trust in order to maximize the trainer's gains. If anything, it is the trainers who end up being sucked into the pet industry's consumerist machine. The trainer's increased affection for the animal may – and often does – result in increased recognition of the dog's needs and, in turn, in increased spending. The recognition that training is a never-ending process – and not an eight-week-long course that is carried out once in a dog's lifetime – results in acknowledgment of the need to spend increasingly more money on the activity (and, as an aside, also fosters the strengthening of community ties among the members of dogdom). Consequently, it is not a coincidence that the American pet dog training community is predominantly middle-class, white (and female): its members need to have the resources necessary to engage in the hobby, as evident in Donna Haraway's analysis of the agility community from *When Species Meet*, where Haraway describes the culture of weekend trips, RVs and expensive physiotherapy appointments (Haraway 2008).

Nicole Shukin identifies the genealogy of feeling power in Foucault's notion of pastoral power: a form of power Foucault traces to the early days of the pastorate of the Church and which he discusses in the College de France lectures on governmentality (2007). In pastoral power, specific techniques are used to influence the will and self-perception of the flock members to make them desire that which the "shepherd" desires. Nevertheless – and this is absolutely key – this is not a cynical form of power; it is not experienced by the shepherd as a manipulation of the subject's desires or as the forming of artificial needs which would only benefit the shepherd. The shepherd genuinely wants the best for his flock. Elsewhere, I have argued for conceptualizing the human-canine relationship in contemporary training as a pastoral relationship with one significant difference: pastoral power was originally used for the governance of multiplicities, but in this case, it can be applied to the individual human-canine relationship (Włodarczyk 2016b). However, this slight difference really does change the dynamics of pastoral power, as will become clear in the next chapter, where the term is applied to the context of a relationship between an actual shepherd and his flock, as presented in the writings of Donald McCaig. The size of the flock, both metaphorical and literal, matters in decisions about its welfare. In the pastoral relationship, the well-being of the individual members of the flock is interdependent. When the flock is a "pack of two," the strong affective ties that bind the flock members develop between the human and the canine. When the flock is a multitude, the well-being of an individual member may be sacrificed in the name of the greater good of the entire flock. In a flock of two, the well-being of the individual member is equivalent to the well-being of the entire flock.

Pastoral power is by no means a solution; it carries with it its own set of problems, particularly connected with mistakenly establishing what truly is good for another being: a task that seems additionally complicated if the relationship is one between members of different species (this is, of course, something that Foucault did not consider: the flock was a metaphor for him) and if we think of the pastoral relationship as a technology of biopolitics. Anand Pandian, in an analysis of pastoral power in an interspecies relationship, writes of the relationship between oxen and their human caretakers in India. Despite the obvious affection they have for their animals, the caretakers tend to impose a certain human moral code on the oxen and hold the ones which conform to the moral code in greater esteem (Pandian 2008). Because they wish for their goals to converge with the goals of their animal charges, the shepherds find ways of imposing their goals on the animals, reinforcing compliance and, as a result; they become more set in their belief about correctly identifying the animals' needs. In other words, they "know better" what it is that the animals want than the animals themselves. Such are the obvious dangers of a relation of power modelled on that between God and his people.

Still, if reinforcement-based training is conceptualized as a technology of pastoral power, it becomes possible to pinpoint two significant changes that this approach brings about: emphasis on the strong affective relationship between human and canine and the assumed permanence of this relationship. Pastoral power, writes Foucault, is a power of beneficence, a power of care and of responsibility (Foucault 2007, 172). It is truly a selfless power in which the "the shepherd (*pasteur*) directs all his care towards others and never towards himself" (173). Because the shepherd is responsible for his flock, he will never abandon it. Indeed, if we think about the changes in practices that accompany the definitions of training as it moves from one intellectual conceptualization to another, it becomes clear that re-thinking training as relationship-building has the very practical side effect of fewer (trained) dogs being relinquished at shelters. There exist quantitative studies that reveal a very strong correlation between the amount of training a dog has received and decreased likelihood of the animal being abandoned (Salman et al. 1998). In the training memoirs I have discussed, it is obvious that giving a problem dog away is not an option. Sometimes, well-meaning onlookers attempt to pressure the narrators into relinquishing their dogs. The authors, however, react quite violently to such suggestions. Garrett writes: "[r]outinely, well-meaning individuals suggested that I should consider placing Buzz in another home (…). I also knew that even if this wild puppy never was able to control himself enough to do agility, he was never going to live in any home but mine" (Garrett 2005, 67). Because it aims at creating an intensive two-directional affective relationship, reinforcement-based training (if engaged in with conviction) is practically a guarantee of a life-long relationship. The title of Karen Pryor's 1984 book – *Don't Shoot The Dog* – caught on among the dog training community because it promised a technology that would ensure just that. Even though Pryor objected to the title, it was actually not misleading; a side effect of the technology Pryor was promoting was a lasting and strong relationship and the end of canine expendability: a frame of mind in which it becomes inconceivable to get rid of one's dog simply because he does not live up to one's expectations. Yet this side effect is not a direct result of behaviorism *per se* – recall Skinner's laboratory pigeons and rats and their lack of a relationship with their caretakers – it is the result of the technology conveniently falling into a community of humans who had long been searching for something like it.

I have elsewhere tried to explain the predominance of women in positive training communities, positing that the redefinition of training as a technology of love and as carework, has made the activity more aligned with dominant cultural constructions of femininity (Włodarczyk 2016b). When we look at the gender breakup of the authors of the positive training textbooks under analysis in this chapter and in the previous one, it is striking that the authors of the positive training books in the 1980s–1990s were almost exclusively female (Morgan Spector and Ian Dunbar being the two exceptions).[6] Meanwhile, in the 1980s, 1990s and into the 2000s the authors of the most popular "traditional" pet dog training textbooks (those not relying on Skinner) continued to be predominantly male: William Koehler (who passed away in 1993), The Monks of New Skete, Matthew Margolis, Milo Pearsall. Philosopher-trainer

6 Trainer Gary Wilkes teamed up with Karen Pryor in the 1990s to promote positive training. The two worked together for some time but broke up their partnership after Wilkes renounced purely positive training and began arguing for a more balanced approach, that is for the necessity of punishment.

Vicki Hearne, whose work is discussed in the next chapter is an exception in this category. Before Blanche Saunders's entry onto the textbook market, there had been hardly any women authors of dog training books in the US.[7] Tellingly, all the textbooks discussed in chapters 1, 2 and 3 were authored by men. Even though the entry of women into the field as trainers and not simply recipients of male trainers' expertise (recall the chapter on "Ladies' Lap Dogs" in S.T. Hammond's book) took place already in the 1930s, it is only with the entry of positive training that women take the field by storm. The quantitative studies of the gender balance of participants in recreational canine events in the United States reveal that activities such as agility and obedience are currently – in the twenty-first century – absolutely dominated by women (Hultsman 2012, Gillespie et al. 2002). Stanley Coren's informal calculations on the prevalence of women in non-competitive events (at companion dog training classes and seminars) estimates their participation at 84% percent. Coren also establishes that 85% of members in the professional organization for Canadian pet dog trainers – specifically, the major organization that favors positive dog training: Canadian Association for Professional Pet Dog Trainers – are women (Coren 2016a).

While there do exist studies that suggest that women may be better trainers because they are more adept at establishing a non-threatening relationship with a dog (cf. Wells and Hepper 1999), these analyses completely miss the point. The question is not: "Are women better trainers than men?" but: "When did we move toward treating the ability to establish a non-threatening relationship as an evaluative criterion for measuring potential success at training?" The answer is: with the entry of positive training, which in turn springs from the need for a technology of love. The groundwork had been laid with the rise of "feeling power" in relation to companion animals, but within the broad field of human-companion dog relations, training had remained a vestige of masculinity. Once we move from seeing training as simply the formation of a docile body; once we move away from the purely disciplinary framework, this is when training begins to be associated with what has been traditionally defined as woman's sphere: it is a labor of love and of care, and of relationship-building, a sphere of emotions. Training becomes redefined as women's work. Even though Nicole Shukin does not discuss gender in the genealogy of feeling power that she traces – in fact, she locates the origins of feeling power in the "masculine" novels of Jack London – it is clear that emotions such as compassion and empathy have long been coded as feminine, and Shukin's genealogy

7 Winifred Strickland was another female author of training books who published at the same time as Blanche Saunders did.

of feeling power in relation to animals could easily be traced back well into the nineteenth century, especially if one focused on so-called women's culture. After all, sentimentalism, which Jennifer Mason in *Civilized Creatures* (2005), saw as fostering feelings of empathy for animals was clearly a feminine aesthetics. While the work of feminist scholars examining the impact of women's values on the public sphere is clearly mixed, they are generally in agreement about the rise, in the second half of the twentieth century and into the twenty-first, of this ethics of care and kindness as a dominant value in popular culture and, increasingly, a guiding principle for certain practices, the most obvious examples being child-rearing and formal education. The cultural "feminization" of an entire profession, that is its re-definition as providing care and requiring love, can be – and usually is – reflected through gradual changes in the gender make-up of its practitioners; veterinary medicine is a particularly interesting example. The percentage of female students of veterinary medicine in the US has increased from less than 40% in the 1980s to almost 80 % in 2007 (Irvine and Vermilya 2010, 62). Companion dog training becomes another field in which the two conjugate processes can be observed: the literal feminization of the profession along with the cultural encoding of the activity of training as feminine.

CHAPTER 6

Countermodernity: Resistance to the Positive Training Revolution, 1980s–2000s

> Thinking back on Kant's text, I wonder whether we may not envisage modernity rather as an attitude than as a period of history. And by "attitude," I mean a mode of relating to contemporary reality; a voluntary choice made by certain people; in the end, a way of thinking and feeling; a way, too, of acting and behaving that at one and the same time marks a relation of belonging and presents itself as a task. A bit, no doubt, like what the Greeks called an ethos. And consequently, rather than seeking to distinguish the "modern era" from the "premodern" or "postmodern," I think it would be more useful to try to find out how the attitude of modernity, ever since its formation, has found itself struggling with attitudes of "countermodernity."
>
> MICHEL FOUCAULT, "WHAT IS ENLIGHTENMENT?"

∴

Not all trainers enthusiastically embraced the exclusive use of positive reinforcement in training in the 1990s and early 2000s, which is when the changes described in the previous two chapters were taking place in the US and Canada. In fact, as the opening section of chapter 5 argued, it is something of a small miracle that Skinner's ideas were accepted by any dog trainers at all. The fact that they were was influenced by the seismic shift away from the classic disciplinary mode in many fields of human-human interaction and social organization: the turn away from corporal punishment and towards motivational techniques in schooling and parenting being the most obvious case in point but certainly not the only example. It is not a coincidence that the disciplinary model in dog training begins to decline as we enter what has been described first by Deleuze (1992) and then by Hardt and Negri (2001) as the society of control, with its dominant model no longer being the factory but now the corporation; with its focus on motivation as a predictor of productivity. Behaviorism seems a perfect methodology for the society of control

 | DOI:10.1163/9789004380295_008

and, as the previous chapter argued, it is the quintessential methodology for the self-formation of the neoliberal subject. Many trainers consciously downplayed these uneasy thoughts as they turned towards reinforcement-based training and embraced it as a "technology of love," believing that it offered liberation from coercion and recognition of the dog's capability to think. In other words, behaviorism became the dominant methodology for shaping human-canine interaction not because of – but despite – its apparent incompatibility with changing ways of experiencing the human-canine bond as a deeply ethical relationship.

Bearing this in mind, resistance to change was inevitable and should not be brushed off, as it sometimes is, as the result of the resisters' general narrow-mindedness and lack of acquaintance with developments in the science of canine behavior and cognition. It deserves to be taken seriously as a response to a certain episteme that the resisters opposed. Analyses of "regressive" training methods – often circulating in electronic form among the dog training community – usually focus on referencing scientific research to expose the myths behind the training methods. What this approach, important as it is, does not take into account are the reasons for the popularity of that which is dubbed "unscientific." Simply debunking myths via science does not reveal the sources of these myths and does not explain the hold they can have on the popular imagination. This chapter analyzes the opposition to the use of positive reinforcement in dog training in the writings of three trainers active in the 1990s and 2000s: Cesar Millan, Vicki Hearne and Donald McCaig. As a philosopher, Hearne was writing for a predominantly academic audience; McCaig – in addition to being a dog trainer – is a well-known novelist whose books are also read by a non-specialist readership (by which I mean readers who are not even dog owners); while Millan, the most widely recognizable name among these, has become a celebrity as a result of hosting his own dog-training reality television show. His ghostwritten books are only a follow-up to his success on the small screen. These trainers come from radically different backgrounds and target very different audiences; yet, I have chosen them as representative of three trends within the "anti-enlightenment community." Even if they promote similar methods, their grounding is quite different. Millan is the anti-intellectual folk expert whose common-sense solutions are an attempt to rescue training from what he perceives as an excess of femininity acquired as a result of the positive turn. Vicki Hearne's opposition to clicker training is an intellectual response to what she sees as behaviorism's misreading of what constitutes coercion, but also a form of resistance to the impeding posthumanist turn which she senses in behaviorist principles (but which she – in counterstance to the argument I have pursued in the previous two chapters – evaluates negatively).

This, in turn, brings her close to Donald McCaig, whose ideas are rooted in a melancholic attempt at recreating a relationship in which the ethical dimension of the human-canine relationship is embedded in a framework of utility and a very clear dislike of technological progress. It seems that what ties these arguments together is a countermodern stance; a reaction against the features of reinforcement-based training that shape it into an enlightenment discourse: its universalism (learning theory works for all organisms), its progressive approach, its rationalism and technophilia.

1 Disciplining Affects: The Dog Whisperer

Cesar Millan has most likely been the most visible and controversial figure on the American dog training scene in the first two decades of the twenty-first century. Millan rose to prominence as a result of the initially surprising popularity of his television show *The Dog Whisperer* (aired on National Geographic between 2004 and 2012, and since 2014 on Nat Geo Wild as *Cesar 911*) but he has also co-authored, with journalist Melissa Jo Peltier, a number of books on the topic of human-canine interaction (to be specific: seven books within a period of ten years). He prefers not to call himself a dog trainer but a dog rehabilitator; finding the word training too firmly associated with the teaching of specific behaviors and – assumedly – the tradition of training for obedience competitions. Millan is not interested in teaching "sit" and "down" but, in dealing with "problem behavior," a term that largely serves as a euphemism for aggression.[1] However, Millan's refusal to call himself a dog trainer also correlates with his focus on the human subject of the show more than with an actual break with the "traditional" methods of dog training. The human subject; a subject whose excessive affects need to be disciplined; is inseparable from the television genre Millan becomes a part of the makeover show; a genre closely linked with themes of self-governmentality of the neo-liberal subject. It has been argued,

1 It is interesting that while in the first seasons of *The Dog Whisperer* Millan attempted to resolve various issues (including fear, separation anxiety, etc.), aggression absolutely takes over the new show, constituting about 80% of the cases Millan tackles in *Cesar 911*. The most popular sections of the episodes are Millan's face-to-face confrontations with aggressive dogs (as evident, for example, in the excerpts of the episodes uploaded to Youtube by fans). Shifting focus to aggression ensures the availability of the physically conformational scenes to the viewers; at the same time allowing for the replaying of what I have called in Chapter 3 the primal scene of dog training: the spectacle of physically enforcing the dog's submission.

for example by Brenda Weber in her analysis of makeover shows that focus on changing the appearance of women's bodies through plastic surgery, that the makeover disciplines the subject's body in a classically Foucauldian way: the body is physically molded to conform to a normative shape (B. Weber 2009). However, as Weber points out as well, the power of the norm is also enforced in more subtle ways; ones that could be read using the more modern paradigm of Foucauldian biopower, which works through the shaping of the subject's desires, through what she calls "affective domination" (Weber 2007, 78). Here is where Foucault's theory of power intersects with Cesar Millan's dog training show twice; firstly, by providing a framework for analyzing the techniques applied by Millan to the dogs' owners; secondly, for structuring reflection on Millan's place in the recent shifts in the functions of training.

I have argued in the previous chapter that the positive training revolution was sparked by "feeling power": the need to acknowledge and shape the affective ties that bind the human and the canine. Millan's methodology sees this attachment as aberrant and in need of being curbed: even though Millan's show overtly pertains to dogs, the covert goal of Millan's rehabilitation program is to curb the human's excessive attachment to the dog: to enforce the power of the norm. Despite the spectacular transformation in the canine's behavior that can be observed in the episodes of his shows, Millan's major premise is that it is the human owner's attitude – "imbalance" or "nervous energy" to use his own vocabulary – that lies at the root of the dog's behavioral issues. If the human's issues are resolved, so will those of the dog. Consequently, there can be discerned in Millan a blatant disregard for the individual personality of the dog, his or her quirks, likes, dislikes and fears; all subsumed within the pack leader mentality according to which the dog simply must do what the human wants, no questions asked.

Secondly, Millan's phenomenal popularity is both a response to and a form of backlash against the rapid growth of reinforcement-based training (sometimes referred to in shorthand as "clicker training"): both the methods Millan promotes and the relationship he strives for are incompatible with the behaviorist approach, which serves as the scientific grounding for these newer methods. The past two decades have also witnessed an immense outburst of interest in the issue of canine behavior (and its modification) among the scientific community. The findings of these scientific studies are something that Millan – ever the folk expert – completely disregards. Not surprisingly, the positive dog training community, canine cognition and learning experts, respected trainers and science journalists have responded to Millan by trying to debunk the myths he propagates: a process easily traced via a number of statements issued by canine-related organizations (e.g. in 2006 by the American Humane

Association),[2] articles in opinion-forming magazines (e.g. Mark Derr's 2006 article in *The New York Times*), on trainers' websites and even social media groups (e.g., there exists an active "Anti Cesar Millan" Facebook Group). The intensity of these efforts, however, points to the fact that Millan has somehow managed to capitalize on trends discernible among the American pet-owning public and, more generally, in American popular culture. His ideas respond to the growing resistance to the mainstreaming of reinforcement-based training, with its roots in the laboratory, its belief in progress and rationality. If we think about Millan's ideas as a backlash to, as Mark Derr puts it, "40 years of progress in understanding and shaping dog behavior and in developing nonpunitive, reward-based training programs," (Derr 2006) it becomes possible to diagnose the nodes of discontent that emerge in response to positive training not just as a method, but as a worldview. The range of Millan's eclectic views reveals the extent of the fears and anxieties that the American public experiences at the turn of the century. These anxieties are evoked not only by clicker training, but clicker training and behaviorism are in this story an element of broader cultural change: an element, really, of the culture wars in which the values of liberal, intellectual America are seen as infringing on the traditional way of life. Millan responds to anxieties evoked by: the redefinition of the role of the expert in animal training (but also, more generally, his popularity is part of the redefinition of the role of expert knowledge); the growing influence of science and empirical research on the practices of dog training (and the growing "scientification" of human and animal life under contemporary biopolitics); the perception of training (and human education) as a never-ending process; the feminization of training (and, more generally, the feminization of American cultural values: the growth of sympathy as an affect binding the American public, as explored theoretically by i.e., Lauren Berlant); the progression of technological developments, which is perceived as deepening the rift between humans and a nostalgically idealized "nature"; and, finally, the progressive blurring of the human-animal divide, aided by developments in science and the rise of posthumanism in the academy, but also discernible in myriads of pop cultural narratives.

Aided by the format of the reality television makeover show – one which Brenda Weber has interpreted as the perfect embodiment of the contradictions inherent in cultural narratives of neoliberal selfhood, citizenship and Americanness – Millan sets off on something of a populist anti-establishment

---

2 American Humane Association, "Dog Whisperer's Approach More Harmful Than Helpful," American Humane Association, September 16 2006, http://www.americanhumane.org/about-us/newsroom/news-releases/06-dog-whisperer.html (Oct. 6 2016).

crusade that allows him to become an expert precisely through his renunciation of expert knowledge. The reality television expert's authority does not result from his academic credentials but, as Tania Lewis puts it in an analysis of the functions of the expert figure in the media, *Smart Living: Lifestyle Media and Popular Expertise*, from his immersion in the "vernacular cultures of the everyday" (T. Lewis 2008, 72). Millan's popularity comes at the time when reinforcement-based training is becoming mainstreamized through its overt recognition by many of the leading animal welfare organizations and outspoken experts on canine behavior and cognition. It also arrives when research on canine cognition and learning is intensifying and starting to transform the field of ethology. All in all, Millan appears when the academia begins to be recognized as the proper place to seek knowledge on dog training. Millan's expertise supposedly comes from an inborn sensitivity to animals: a gift, so to speak, that he possesses despite his lack of formal education. This is nothing new in the world of animal training; as John Homans put it, "mystics have always had a place in the dog world, communicators with the hidden worlds of nature." (Homans 2012, 44). In fact, one of the innovations that came along with behaviorism was the understanding of training as a mechanical skill – this is pioneer behaviorist Bob Bailey's long-term adage – that can be honed as a result of both theoretical study and years of practice. Millan seems to counter these beliefs in a way that resonates with the general public by speaking of the disconnection between one's education and one's ability to navigate "the real world." In fact, according to Millan, too great an emphasis on theoretical knowledge leads to an "imbalance" in the person: one which is sensed by the animal who refuses obedience to the imbalanced human. In a 2010 interview Millan states:

> Many people are trained, meaning they have a degree, but they're not balanced, right? Someone might be a Harvard graduate, but can't walk a chihuahua. That human is trained, but doesn't know how to employ a more simplistic way of relating. The dog only sees, okay, you're not balanced, therefore my integrity doesn't allow me to follow you (. ...)
>
> You can be a Ph.D., an M.D., and have whatever Masters you want to have – the dog is not impressed by it. You can have the most expensive suit in the world – a dog is not impressed by it. You can be the most famous human in the world, an actress, an actor – the dog is not impressed by it. (Millan quoted in Cochran 2010)

The value placed on simplicity and balance, traits associated both with some form of the mythologized past and with the essence of dogness, respond to the takeover of dog training by academic experts: the ever-multiplying numbers of publications for solving specific behavioral problems through the implementation of complex and detailed protocols of counter-conditioning and desensitization. Millan also responds to the growing subjectivation of the dog in training protocols: the progressive blurring of the canine-human boundary that can be detected both in behaviorism's universalizing principles (all organisms learn according to the same principles) but also in the sensitivity to the dog as an individual being that accounts for the complexity of the long behavior modification protocols: the belief that Sparky may require a different procedure than Rover to solve the same behavioral problem. Millan promises a simple fix to all problems: a common-sense approach, a form of folk knowledge that, as he suggests, gets lost along with the behaviorists' hostile takeover of the field. Millan's training methods do not form a coherent methodology, which is why it is next to impossible to attempt to summarize them as such, even though some scholars have attempted to point out the specific inconsistencies in his approach. These include Millan's views on the structure of "the pack," the number of potential leaders, transference of leadership, etc. (Jackson-Schebetta 2009, Pręgowski 2015). Millan is not consciously responding to a certain tradition of training and trying to position himself in relation to specific schools. There are influences obvious to the eye of those familiar with the history of dog training (Most-Weber-Koehler-Monks of New Skete), but these are not mentioned. Millan does not feel he needs to explain himself as operating in relation to anyone: appropriating multiple traditions without any footnotes, he claims to be inventing something radically new and yet falling back upon an ancient tradition of shamans possessing special powers of communicating with animal spirits.

Millan's words, interestingly, speak to a public that has already redefined the human-domestic pet relationship as compensation for a certain loss, lack of "nature" in their (sub)urban middle-class lives. For this public, the paradigm of civilizational progress that behaviorism has insisted upon – we move from the dark ages of traditional training to the Enlightenment of clicker training – carries with it a certain ambivalence related to a growing consciousness of the dark sides of civilization and the omnipresence of expert knowledge. I have said that Millan speaks "to the public" and not just to dog trainers and this is most certainly the case. In the early twenty-first century, the market of dog training publications becomes increasingly specialized: training books reference other books and focus on solving specific behavioral issues or teaching individual skills. Paradoxically, less information is available for the new dog owner who is not seeking to enter a debate on dog training but wants an easy fix to the issues he experiences with his or her own dog. Through his unprecedented use of television during a time when the reality TV format is at its height, Millan becomes the first point of reference for many new dog owners, which also means that the themes he picks up on: the need for simplicity, renunciation of a purely scientific approach and the idea of "reconnecting with nature" resonate not only with members of the dog training community but form the undercurrent of American popular culture. There is, of course, an obvious paradox in what Millan is proposing: book knowledge is nothing, you need a gift of speaking to animals (he actually uses the Dr. Dolittle analogy) to communicate with them. It would seem that if this sensitivity is inborn, then it cannot be acquired by the average pet owner who has, as Millan claims, lost touch with nature. Yet Millan's role as expert is to translate animal language in a way that makes communication possible, even to American owners who have been evidently impaired through their immersion in civilization. Part of the appeal of Millan is the – even clearly hypothetical – potential of interacting with Millan that his show projects: the Dog Whisperer enters one's life and changes everything for the better by modifying the behavior of the family dog. Millan becomes the type of expert representative of makeover culture: the lifestyle expert.

In his diagnosis of the roots of the dog's misbehavior – the diagnosis is constant: a lack of what Millan calls calm, assertive energy in the dog's human owner – Millan locates the roots of other problems as well: problems in the human's interpersonal relationships and even problems related to professional success. Adopting his solution, which basically consists of developing calm assertiveness, is a fix for all problems in the life of the human subjects; not just those related to their pets. Scholars of reality television have been invariably emphasizing that the subject of reality television is the self as a site of potential transformation (T. Lewis 2008; G. Palmer 2008; B. Weber 2009). Even if the overt theme of a particular show is the participant's house or car, the subject is educated on the behaviors expected of a proper citizen of the neoliberal state he or she inhabits (Oulette 2009, 223). Conversely, the tips Millan prescribes as a solution to the dog's problem behaviors are solutions that are meant to be implemented by the (often female) owner: exhibit more calm assertive energy and all will be well. Consequently, the fact that Millan's show pertains to dogs is of secondary importance because the main recipient of his teachings, and the true subject of the show, is the dog's owner. This explains the main paradox of the show, as pointed out by Lisa Jackson-Schebetta; despite Millan's emphasis on canine difference (with the resulting need for the human to assume the pack leader position) and his resistance to anthropomorphism, the dogs at the end of the show are cookie-cutter images of one another; products that conform to modern anthropocentric ideas about what the dog's role is: to be friendly, cuddly and non-aggressive; non-disruptive to the lives of his human caretakers (Jackson-Schebetta 2009, 107).

Lifestyle experts are not traditional experts whose knowledge is backed up by professional credentials. Millan has a gift, or so he shapes his life narrative. Yet what makes Millan an expert is not just his inborn gift but the trajectory of his life in which he manages to capitalize on this gift. Consequently, he becomes the perfect embodiment of neo-liberal selfhood, with its promise of upward social mobility firmly tied to individual entrepreneurship. The Dog Whisperer is a lifestyle expert because he has managed to fashion his life in an expert manner: he has overcome hardship and poverty and has lived to tell the story from the vantage point of fame and wealth. He is the Benjamin Franklin of reality television, but he also needs the published word to tell his story as such, and his books complement the show by presenting that which the reality television format necessarily omits: Millan's pre-television life. *Cesar's Way*, Millan's first published book, is largely autobiographical in the classic Enlightenment definition of the genre: it is a story of someone who manages to become successful through his own hard work and entrepreneurial spirit; referencing Benjamin Franklin's autobiography and Horatio Alger's "rags to riches" narratives also through the necessary severing of familial and community ties as a result of an escape at a formative age: Millan runs away from the poverty of Mexico and illegally enters the US at the age of twenty-one. His dream, as it is presented in the story, is to become a dog trainer, but there are hints throughout the book that this dream may have been imposed on Millan's life narrative *ex post-facto*. As Millan reflects on why he accepted a job offer made by the owner of a limousine fleet – "he was a very impressive man – the kind of strong, confident businessman I wanted to become someday" (Millan and Peltier 2007, 48) – cracks begin to be visible in the official narrative of "the dog whisperer." Millan simply wanted "to make it" in Hollywood and correctly identified a niche that ensured his success. Lisa Jackson Schebetta writes of Millan as "enacting his own version of the American Dream." (2009, 123). Yet, paradoxically, it is through fashioning his story as the ultimate American "rags to riches" narrative that Millan gains the license to criticize Americans: on the one hand, he shares the dreams of millions of Americans; on the other, his wisdom comes from his success at fulfilling the dream (as opposed to their failure).

Millan is the perfect enterprising American citizen, playing up his ethnic and class heritage in a way that makes it possible for him to succeed not despite but because of his otherness. He does not refute the American perception of the dark other as somehow closer to nature but learns to profit from it. He consciously fashions himself as a "wild child," emphasizing the time spent on his grandfather's farm, where he observed the ranch dogs interacting with one another through displays of "calm submission" and "active submission." (Millan and Peltier 2007, 27). Yet, an attempt at creating a timeline out of the events presented by Millan in *Cesar's Way* reveals that he never actually lived in the countryside. Millan was born in a city with a population of approximately 600,000 (Culiacan) and while he did visit his grandfather's farm on weekends and vacations, these visits ceased, or at least decreased in frequency, when his family moved to another city, farther away from the farm, in search of better educational opportunities for the children. This move happened when Millan was "six or seven years old." (29). In other words, a close reading of Millan's own account shows that The Dog Whisperer never actually lived outside of the city, and the period to which he refers as his formative years as an observer of the pack life of rural dogs occurred during weekends and holidays and ended when he was six years of age. Still, he skillfully uses Americans' stereotypes of Mexicans to market an image of "the noble savage" who has not been corrupted by civilization and is thus capable of reflecting on its dangers. He delivers exactly the kind of reflection that an urban American expects from him: "What we seem to forget is that we still have access to the very same world they [animals] inhabit. That's why indigenous peoples in the deserts, in the mountains, in the jungles are able to survive for generation after generation. They are Homo sapiens (sic) just as we are, yet they are fully in tune with their animal natures" (270).

Millan is also very skillful at maximizing the potential associated with his gender. Lewis links the rise of the "lifestyle expert" with a change in the status of what has traditionally been defined as the domestic domain of women: "a repackaging of feminine expertise under the sign of 'the lifestyle expert.' " (T. Lewis 72). The case of *The Dog Whisperer* is somewhat different: dog training in the early twentieth century had been quite firmly associated with rugged masculinity. It is only toward the end of the twentieth century that the field of dog training begins to be taken over by women, as training becomes redefined as the "technology of love." The change from the traditional disciplinary methods – understood in the classically Foucauldian sense: methods that focus on the body by molding it into docility – to the more modern methods that affect the dog's desire to cooperate with the human is also tied to a gender change. As training begins to be seen as carework and not the taming of a wild beast, the activity becomes associated with femininity.

The entry of hypermasculine Millan, with his traditional methods, rescues dog training from the excess of feminine emotions that have implicitly ruined the effectiveness of the practice. The main problem that Millan diagnoses with American pet owners is that they love their animals too much and view their relationships as channels for expressing affection (Millan and Peltier 2007, 55–57). Even though Millan does not speak of this excess of affection as specifically a women's issue, the gender breakdown of the pet owners in his television programs is definitely uneven. While most of the dogs are family pets, and Millan makes sure that all family members interact with the dog, it is often the women in the families who are their primary caretakers. Millan thus often assumes the well-known role of the male expert (with examples ranging from Sigmund Freud and Dr. Spock to Mr. Clean) whose intervention in a woman's domestic life changes it for the better; or at least would change it for the better if she followed Cesar's rules diligently. Some of the women do not, and it is assumed that their lack of success results from their inability to follow Millan's instructions. Consequently, as viewers, we are invited to become exasperated with the lack of consistency exhibited by the female owners. Ashley E. Pryor writes of an episode that centers around the resolution of Punkin's (the dog's) rock obsession, which Millan manages to solve, but which the dog's owner, Amanda, still cannot control once Millan is out of her house: "Not surprisingly, while Punkin is able to shift his attention away from rocks to engage in a social relationship with his stable pack leader Millan, he reverts to his old fixation when he is reunited with the energetically imbalanced Amanda" (A.E. Pryor 2012, 302).[3] Within the framework of the show Millan's balance – his

3 Ashley E. Pryor, "Heidegger and the Dog Whisperer: Imagining Interspecies Kindness," in *Animals and the Human Imagination: A Companion to Animal Studies*, eds. Aaron Gross and Anne Vallely (New York: Columbia University Press, 2012), 289–306. Ashley E. Pryor's article is an example of the positive reception of Cesar Millan by some animal studies scholars: something I initially found highly surprising, because of the significant overlap between animal studies scholars and animal rights activists. Pryor argues that Millan's calm-assertive energy is an example of Heideggerian kindness (particularly as opposed to the hysterical and aggressive energy exhibited by the dog's owner). In brief, Pryor claims that it is kinder to issue one strong correction to resolve the problem behavior once and for all rather than to focus on a potentially futile search for positive solutions. Conversely, Julie A. Smith, in the volume *Experiencing Animal Minds* (2012), analyses Millan's interactions with the dogs in the episodes of his show through the notion of KVT: kinesthetic visual transfer, or the ability to experience sensation in oneself while observing it in another: a transfer of bodily sensation as a result of its visual observation; a concept that resembles some definitions of intersubjective empathy. Not only does Smith see Millan as possessing KVT (a thesis I would most certainly disagree with), she also argues that Millan's method is based on the assumption that dogs experience KVT and can learn through it. Ultimately, it is difficult to not read the article as interpreting Millan's ideas as progressive because Millan's methods acknowledge that dogs are capable of KVT. I am absolutely not arguing that dogs do not experience KVT; however, what worries me in these two articles (Pryor's and Smith's) is the assumption of the neutrality of Millan's position as well as the assumption of the transparency of his methods as presented in the reality television

calm-assertive energy, which can basically be translated as self-confidence – are tied both to his masculinity and his ethnicity.

In a society of control – a term first introduced in Gilles Deleuze's reflections on the shift away from the Foucauldian disciplinary society – reality television becomes yet another example of the deinstitutionalization of education and, somewhat paradoxically, Millan can be seen as participating in this shift, particularly through his use of non-standard (at least for dog training) channels of knowledge distribution. The popularity of reality television is itself tied to the emergence of new forms of governmentality that work "from a distance," shaping the subject while avoiding direct physical contact. The education of canines ceases to be the province of classic disciplinary institutions, such as the dog obedience school, as the expert gains entrance into the domestic sphere and propagates his techniques via television. Millan's "rehabilitation" is actually a word that reveals his investment in neo-liberal selfhood: with connotations of the subject "letting himself go," succumbing to some form of addiction. While Millan speaks of rehabilitating the dog and training the owner, in an attempt at inverting the traditional use of the way these terms have operated in animal training, it is the human subject of the show who must undergo "rehab," and only his or her rehabilitation is a guarantee of the permanence of the change

---

format. Both Pryor and Smith approach Millan as they would a scientific article published in a peer-reviewed journal, not as a skillfully constructed visual piece of entertainment. Smith assumes Millan's concept of energy transfer is correct because she sees it working in the show. Both Smith and Pryor completely disregard the existence of a "scaffolding" to this show; the editing and piecing together of fragments that remains obscured in the finalized product. Finally, they both disregard the larger framework of the history of dog training methods, seeing Millan as on the one hand somehow representative of dog trainers in general and, on the other, as innovative in his methods and progressive in his approach to canine otherness. This is simply not the case. Millan's methods trace back directly through the lineage of Konrad Most-William Koehler-Monks of New Skete, etc. While the use of the notion of "energy" as the grounding of his practices is indeed new (although the Monks of New Skete also use a similar terminology while insisting on the pack model), the notion of "pressure," routinely used in the training of sheepdogs since at least the early decades of the twentieth century, is somewhat analogous to Millan's energy. In summary, I diagnose the surprisingly positive reception of Millan as tied to the entanglement of animal studies in the counter-linguistic turn (Weil 2012) through the desire to seek examples of meaningful interspecies relationships that are not mediated via language. This possibility of communicating with animals as a result of some special feature particular to the individual seems to hold sway over a group of scholars devoted to understanding the animal mind who generally welcome "communicators" who claim their access to animals is somehow more direct than that of the general population. This fascination with nonverbal communication also explains, at least partially, the generally positive reception of, e.g., Temple Grandin – a scholar who claims her autism helps her see the world through a non-human's eyes – among some animal studies scholars.

in the dog's behavior. The "rehab" consists of learning to curtail one's excessive affection for the dog. In the example above, Amanda's lack of success at exhibiting "calm-assertive" energy results in problems with Punkin's behavior. In the case of the subjects of Millan's shows, the owners are guilty of undisciplined affects; of allowing themselves to love their dogs too much and unwisely.

The dynamics of the process suggested above can easily be traced through a close analysis of specific episodes of Millan's shows. Here, I will briefly look at two episodes (Episode 1 and Episode 5 of Season 1) of Millan's more recent series: *Cesar 911*. Both of the episodes deal with canine aggression, and in both the root cause of the aggression is diagnosed by Millan as an excess of love that results in the human's inability to structure the human-canine relationship. Additionally, in both cases, excessive love for the dog clouds the (female) human's evaluation of what should matter the most: relationships with other humans. The two episodes are also highly exemplary of Millan's methods: in both cases there is very little actual dog training involved. It is the owners of the dogs who are forced to confront their (supposed) emotional issues, which are diagnosed as underlying the dog's misbehavior. The subjects inevitably break down crying once they understand how guilty they are of letting their runaway love for their dogs ruin not only their dog's lives but also their interpersonal relationships with other family members and friends. In each episode of the show Cesar Millan takes on – and, inevitably, successfully resolves – two cases, which are brought to his attention by so-called "whistle blowers": friends of family members of the struggling dog owner, who contact Millan in what they view as a last-ditch attempt to help the owners deal with the dog's problem behaviors.

One of the two cases in the first episode, titled "Family Feud," involves a medium-sized mixed-breed dog named Auggie, who exhibits aggressive behavior towards other dogs. The situation is causing strife between two adult human siblings: Auggie's owner Nicole and her "whistleblower" brother Brad, who claims Auggie's aggression towards Brad's dogs – Kemo and Sophie – is making it impossible for the siblings to pursue their familial bond. Brad describes the situation to Cesar by theorizing that "Auggie is anxious because Nicole is anxious," and Nicole's husband Matt seconds Brad's diagnosis. Brad's wife adds that they are all a very close-knit family who like to have big family events together, and the issue with Nicole's dog is making that next to impossible. It is interesting that even though Auggie is the family dog, his behavior is directly diagnosed as resulting from Nicole's problems: both by the men of the family and by Cesar Millan. After briefly observing Nicole and Auggie outside on a walk, Cesar determines that Auggie's aggression issues are a direct result of Nicole's inability to trust her dog around others: "Nicole is classic, classic neurotic dog owner. (…) She doesn't trust her own dog to be friendly (…). It was absolutely Nicole's lack of trust that causes Auggie to bite. If you panic, you become a source of what they should bite because you're in a weak state." Cesar's solution is to take Auggie, Nicole and Matt to the dog park, let the dog off leash and observe Nicole's behavior. While Nicole is, indeed, anxious that Auggie might bite the other dogs in the park, Cesar explains to her that it is her own weakness and imbalance that makes Auggie attack. If only she remains calm, the dog will not misbehave. Despite Cesar's admonitions, Nicole tenses up as Auggie starts sniffing other dogs. Consequently, Millan asks Nicole to leave the dog park, as punishment for her inability to remain calm. The technique he uses on Nicole – usually referred to as the "time-out" – is often associated with disciplining unruly children and has been advocated by another reality television expert, the Supernanny. However, it is also a technique firmly entrenched in positive dog training methodology; a fact that has made its way into the realm of high theory. In fact, in *When Species Meet*, Donna Haraway describes using the time-out on her dog Cayenne to teach the dog not to jump over the contact zone on the A-frame, an obstacle on an agility course (Haraway 2008, 205–248). Here, however, the situation is reversed. It is not the dog who is subjected to the time-out but the dog's owner.

Nicole leaves the dog enclosure and breaks down in tears, realizing that she has – yet again – been guilty of letting her improper emotions influence her dog's behavior. The sobbing Nicole is taunted by the three men – Cesar, Brad and Matt – and by Brad's wife, who jokingly speak of putting Nicole "in quarantine." Finally, once she composes herself, she is allowed back in by Brad who asks: "Do you want to come back and play with us? Just be social." Nicole does come back into the dog park and is asked to walk around the park with Kemo, while remaining calm. She struggles with the task but manages to pull it off. In the end, when Auggie does not misbehave, Nicole exclaims "This is like a new dog!" and thanks Cesar profusely. While it is of secondary importance here whether and why Cesar's techniques actually work – after all; how can one evaluate Nicole's calmness? There is a circular logic in place here: the dog's behavior is seen as proof of the human's mindset, Nicole is only diagnosed as calm as long as the dog remains calm – what is of primary importance is that Millan is fully successful at convincing others of his diagnosis: the woman's excessive love for the dog is at the root of the problem, and the issue becomes a problem only because it interferes in family life: Nicole cannot participate in human social functions effectively. Her relationship with her dog trumps her relationship with her family and this must change. It is only after Nicole agrees to "be social" that she is allowed to reconnect with Auggie but on different terms.

One of the two cases in Episode 5 of Season 1 of *Cesar 911*, titled "Ripped Apart," contributes to the analysis above by showing, once more, that runaway love is not only at the root of the female subject's inability to smoothly function with her dog but also ruins her romantic relationship(s). In this episode, Aileen, who has been involved in dog rescue for years, is shown as struggling with Titan, a pit bull that she adopted off the streets. The dog's problem is described as aggression, although the only display of aggression we actually witness during the show is the dog's snarling when Aileen attempts to open the crate in which Titan is locked. The whistleblower on Aileen is her boyfriend Brent, who is exasperated with the strain the dog's behavior is putting on their relationship. In fact, Brent threatens to leave Aileen if she does not solve Titan's issues: "So much of the contention in our relationship can get tied back to the dog. She made the decision without me, and now his presence has been a major factor in there being a ton of fighting between two people who were very happy." Aileen's caring and compassionate nature is presented as being excessively so: her runaway love for animals has made her jump to a decision, without first consulting her boyfriend. After witnessing the dog's behavior on a walk, Millan diagnoses the problem as Aileen's anxiety. He speaks of her being incompatible with the dog because of Aileen's "anxious and impatient ways." Aileen tearfully admits to feeling angry both with Titan and with Brent and agrees that she wants her family to feel like a family again. To resolve the issue, Cesar decides to take Titan into his training facility – which he refers to as the Dog Psychology Center – an act that serves as a form of punishment for Aileen, who "misses him terribly," and is worried Titan will never be returned to her. As Aileen and Brent visit Titan, Aileen says: "I'm really excited to see Titan, but I'm also worried that Cesar might not think we're ready to take him home."

Slowly, Aileen begins to understand that, in order to gain Titan back, she is going to have to learn to control her emotions: particularly her anxiety, anger and impatience. Tellingly, the only times she exhibits these "excessive" emotions occur during her interactions with her boyfriend; not with the dog. As the couple interacts with Titan in Millan's facility, the tension between the humans escalates. During the next visit, it is revealed that Brent and Aileen have split up. Despite this drastic change in the domestic situation, Aileen still wants to gain Titan back. She understands that she can only do this by changing the kind of person she is: "I feel like if I can't change, it's selfish to keep him because I'm going to have an unhappy, unhealthy, unbalanced dog. This experience made me realize that I'm not who I used to be and I want to get back there." Millan's work with Titan consists of teaching Aileen to be "calm and patient" by going on long walks with his pack of dogs. After a while, she is allowed to take the dog back, although only under certain conditions: Cesar installs security cameras in her apartment to monitor her interactions with Titan. The narrative of the show implies that it is Aileen who is to blame for the breakup of her and Brent's relationship: specifically, it is her inability to control her excessively feminine emotions. Her love is not only excessive; it is also inappropriately directed: at the dog and not at the boyfriend. The framework suggests that while her relationship with Brent is beyond salvation, the training she has received from Millan should make it possible for her to enter into a healthy heterosexual relationship in the future.

It should be added that sometimes the subjects whose behavior is marked as excessively loving are male. However, they are diagnosed as exhibiting an inappropriate – and simultaneously "unnatural" – version of masculinity. In Episode 2 of Season 1, "Boss with a Bark," the male co-owner of a Greek neighborhood restaurant, the extravagant and somewhat dandyish Takis, is blamed for the misbehavior of his three fluffy little dogs. Takis dotes on his Bichons, which causes the discomfort of both his wife and his customers, because the spoiled dogs sleep in the couple's bed while at home, and bark at customers while at the restaurant. In fact, the family business is in danger because the dogs are scaring off the customers. Takis needs to learn to act like a man and, as his wife tells him, "be more aggressive." Finally, she openly threatens to "cut off the sex" if he does not kick the dogs out of their bed. The last threat is meted out as something of a last-ditch attempt and has the immediate effect of Takis getting up and sternly forcing the dogs off the furniture. In brief, regardless of the gender of the owner, excessive emotional investment in the dog(s) is shown as tied to excessive femininity or an effeminate masculinity. The implementation of a program of "tough love" is repeatedly presented as the key to success in personal life; and sometimes even in business, as in the case of Takis's restaurant. Cesar's interventions are aimed at restoring the supposedly "natural" order: traditional gender relations are paired with an equally traditional understanding of the dog's place in the multispecies pack. Within the framework of the show, the dog's misbehavior is a result of infringements on this natural relationship and thus, the re-establishment of the human owner's proper mindset will inevitably lead to a change in the dog's behavior. Work with the dog is almost secondary to the work Millan carries out on the human. Yes, the accusations about the roughness of Cesar's methods are true – he does kick dogs from time to time and most certainly uses strong leash corrections – but these techniques are secondary to, and result from, Millan's overarching visions of the human-animal hierarchy, which needs to be restored through disciplining the human's unruly affects. In fact, more often than not, the jerks and kicks are also a way of affecting the dog's owner, who cringes in pain watching the dog being jerked, but grows to understand that tough measures are necessary in order to accomplish change. However, sometimes – as in the case of Auggie and Nicole – there is absolutely no work with the dog. All that happens to Auggie is that he is let off leash in a dog park, while Nicole learns to work on her "calm, assertive energy."

Millan admonishes: " 'Hold back affection? It's not natural!' Please don't misunderstand me. Love is a beautiful thing, and one of the most important gifts we can share with our dogs. But it's not the most important thing they need" (Millan and Peltier 2007, 151). This tendency to love too much is, of course, associated with femininity but also with the effects of civilization: this is a first-world problem, as evident from Millan's early diagnosis of the dogs he meets upon his arrival in California. Effeminacy and overcivilization go hand in hand. Millan's diagnosis brings to mind Lauren Berlant's understanding of U.S citizenship as defined via a shared sense of being "traumatized by some aspect of life in the United States" (Berlant 1997, 1). Berlant's notion of the intimate public, of citizens bound by some private trauma, has been picked up by scholars of reality television who recognize that such trauma is "the reality genre's stock in trade." (McCarthy 2007, 19). In *The Dog Whisperer*, all of Millan's human subjects have been traumatized by the supposedly mundane experience of pet ownership gone wrong. The solution, as it is in all programs of this format, is to step up and change one's life: introduce discipline, curtail excessive affects, get rid of the surplus of emotion that caused this mess in the first place. While such advice is indeed the reality genre's stock trade, there is hardly any place for a respectful understanding of canine individuality within it. There is also no desire to understand the actual nuances of canine behavior modification that exceed the "attitude adjustment" that Millan proposes. Additionally, it is obvious that Millan's show is primarily about Millan; his uniqueness and his special power of communicating with animals. While the tradition of animal communicators is quite long, Millan is the only "communicator" who has used his ability as a springboard to celebrity status. Still, with a show that is (necessarily) so focused on emphasizing Millan's special abilities, one would be hard pressed to expect techniques that are specific and replicable, yet at the same time individually tailored to particular dogs. Millan has built his career around a lack of specificity and the power of presence: his own. His program on Nat Geo Wild is actually preceded with a "Don't Try This at Home" warning.

On the other hand, regardless of how "counter-modern" Millan is in his rejection of scientific knowledge, he does reflect the very modern idea that one of the primary purposes of pet-keeping in the twenty-first century is to "reconnect with nature." He writes: "In America and in some other cultures of the world, dogs and the other animals we bring into our homes serve as one of our most important links to Mother Nature" (Millan and Peltier 2007, 270).

What Millan's popularity brings to the surface is the idea that the seemingly progressive desire to "reconnect with nature" through pet-ownership can, in many situations, be a fantasy that has nothing to do with a deep interest in the specificity of the non-human other with whom one shares one's house. It can be an empty slogan. This can be discerned in the ease with which people fall for the absurd and scientifically disproven theories that Millan is openly advocating: they yearn for solutions that help maintain myths, rather than challenge them. According to Millan's diagnosis, the real reason why Americans have pets is because it makes them feel good to share their lives with a non-human being; though only if it does not disrupt their lives too much. If one of the reasons why Americans keep dogs is to feel good about themselves and experience a link to nature, the methodology proposed by Millan is actually quite coherent and sensible. The dog is an instrument to humans' sense of fulfillment (the title of the last chapter of *Cesar's Way* is "Fulfilling Our Dogs, Fulfilling Ourselves") and if reconnecting with an idealized primal vision of nature through the dog is the goal, then Millan most certainly delivers: he gives us a highly mythologized if completely unscientific version of the human-canine relationship. Most importantly though, Millan's role as an expert is not limited to advice about managing the unruly behavior of companion animals: his main focus is the behavior of humans, which he molds and shapes according to dominant versions of neoliberal selfhood, with a special emphasis on smoothing out wrinkles in the performance of gender roles. Women are at risk of an excess of tenderness that needs to be constrained, while men need to be directed in a better – less effeminate – performance of masculinity

2 Vicki Hearne: On the Nature of Freedom

I purposefully employ the word "moral" as opposed to "ethical" in discussing the work of Vicki Hearne, not only to distinguish from Foucaldian ethics that served as the foundation of my argument in the previous chapter, but because Hearne herself insisted on the moral character of the human-animal relationship. The word moral, with all it religious connotations, is absolutely key for Hearne's project as evident in the title of her most popular book: *Adam's Task* (1986). In fact, she described her ideas as opposed to the two discourses that she saw as dishonest in their refusal to apply the category of moral understanding to companion animals: philosophy and behaviorism. Philosophy, as Hearne read it, rejected animals' capacities for moral understanding and behaviorism reduced all relationships to contingencies of reinforcement. Hearne wrote of animals as capable of "a complex and delicate (though not infallible) moral understanding that is so inextricably a function of their relationships with human beings" (Hearne 1986, 8). Hearne's high opinion of animals' moral capacities, a theme enthusiastically picked up by many contemporary animal studies scholars, made her something of a precursor of the animal turn. However, her opposition to the notion of animal rights as well as her disdain for the discourse of science in general (and behaviorism in particular) account for her difference from most contemporary animal studies scholarship, where science and the humanities form something of a symbiotic and politically engaged relationship.

Vicki Hearne's ideas – most famously presented in the book *Adam's Task* (1986) and the essay "What's Wrong with Animal Rights?" (1991, later included in *Animal Happiness*) – are also memorable among animal studies scholars for her definition of training as a process of developing a language between humans and domestic animals (Haraway 2003; Patton 2003; Wolfe 2003b; Fudge 2008, Weil 2012). Hearne also speaks of training as a relationship that draws both subjects closer together, attunes them to each other and allows for subtle and sophisticated interactions – which she insists on calling linguistic – and which would not have been possible without this common history of working together on a certain goal. These ideas are very similar to – though predating – Donna Haraway's understanding of training from *Companion Species Manifesto*; a fact Haraway dutifully acknowledges in a chapter that pays homage to the memory of Vicki Hearne. It is the human's involvement with the animal that creates and deepens the bond between them: a bond that Hearne views as one steeped in mutual responsibility. Hearne's is a relational ethics, a fact that makes its value questionable as a blueprint for human-animal relations in general (Wolfe 2003b), but relationality is actually what brings her close to crossover trainers whose writings were discussed earlier: it was the specificity of the training bond with their individual dog that made them reflect on dog training in general. It seems that Hearne's ideas are an expression, in a vocabulary more sophisticated and grounded in philosophy, of the same ethical impulse that was behind the discomfort experienced by the crossover trainers. It was the growing importance of the notion of human-canine bonding, the fantasy of "feeling power," that led them to search for a training methodology that did not rely on physical corrections, which they saw as incompatible with the bond they desired. The problem is: Hearne does not experience this discomfort. On the contrary: the methods she advocates, those of William Koehler and his son Richard (Dick), are among the most aversive and physically confrontational in the history of dog training. It is difficult not to question how it was possible for Hearne to speak so beautifully of training as a tool of interspecies communication and proceed to describe how she filled a hole that her dog Salty had dug in the yard with water and then stuck the dog's head in it until Salty choked because she could not breathe. There is a clash, a contradiction, that almost everyone who has ever written about Hearne clearly notices and yet to which no one has attempted to bring some form of cohesion. It seems that there exists a tacit agreement that Hearne meant well, but her personal friendship with Dick Koehler corrupted her estimation of what counts as cruelty. Haraway concludes that Hearne's vision of canine-human happiness constituted "in the labor of training" is an example of "emergent naturecultures" (Haraway 2003, 52). I dare to disagree.

While Hearne's general conclusion that both animals' and humans' (moral) capacities are, at least to an extent, the product of interspecies interactions and are not faculties that exist *a priori* can seem promising when set against certain concepts that have become prominent in the posthuman turn – in addition to Haraway's "becoming with" these include Karen Barad's notions of intra-actions (2003), Birke's performativity of species (2004) or Hayles's relationality (2002) – there are several major flaws in Hearne's thinking that make it impossible to accept her as a precursor of posthumanist naturecultures: her radical humanism (under the guise of anthropocentrism), her regressive reading of nature as immutable and God-given and her total rejection of the discourse of science. It is not simply an unfortunate error that Hearne ends up promoting training methods that are nothing short of abusive. In fact, Hearne becomes an emblem of what can go wrong when the non-human turn chooses not to engage the discourses of science. I do not diagnose the problem as simply being critical of science but as *a priori* refusing to engage with it, which is what Hearne chose to do.

A significant element of Hearne's vision, and much of the resistance to reinforcement-based training expressed in the 1990s, results from the refusal to move away from a very traditional, almost theological, understanding of nature. Beginning with the title of her book, *Adam's Task*, Hearne's vision is a vision of nature as God had intended it: dominion over animals is man's responsibility: once you name an animal, you become responsible for that particular animal (Hearne 1986, 47). The coherence of Hearne's theory rests on this anthropocentric and at the same time biblical assumption of man's dominion over every living creature. The dog's total freedom is thus always illusory because the dog is always under the control of man. The dog's condition is, in fact, a reflection of the condition of man whose freedom is also contingent on obedience to God. Surprisingly, Hearne's diagnosis of the situation of companion animals is somewhat analogous to Skinner's diagnosis of all living organisms. According to Hearne, dogs are never free anyway, thus the idea of giving them freedom by leaving them alone, not asking them to engage in interactions with humans, is fallacious from the beginning: "It won't do to suggest that the dog can just live peacefully around the house while we refrain from giving any commands that might deprive him of his 'freedom,' for that simply doesn't happen. We are in charge already, like it or not" (49). Skinner would most likely concur and say that the organism is always acted upon by the environment, which shapes the behavior of the organism and in the human-canine relationship it is the human who controls the dog's environment. Thus, the idea that the dog can "be left alone" is mistaken from the start because the organism is always being acted upon even if he or she is not conscious of this fact. Where Hearne differs from Skinner is in the understanding of the freedom that is still available under these circumstances. For Skinner, the only tangible definition of freedom was freedom from aversive circumstances; freedom from pain. It so happened that his experiments proved that refraining from punishment was also the most effective way of controlling the behavior of others. Skinner's world is a world of pervasive and constant environmental control and behavioral self-management. Hearne imagines a world where a certain degree of freedom, understood as the possibility of acting in a way that is not determined by others, is possible once total obedience to a set of "ground rules" has been established. These ground rules have to be established firmly and through punishment, which – after Koehler – Hearne calls corrections. Erica Fudge reads Hearne's training as enacting a certain paradox; through corrections it makes possible the transcendence of dominion: "Training is more than a re-enactment of dominion" (Fudge 2008, 94).

Hearne very clearly refuses the dogma of positive training – described in the previous chapter – that is the assumption that the elimination of physical pain is the starting point of choice, what Foucault called in *The Birth of Biopower* an irreducible. For positive trainers, choices in which one of the options is pain and the other avoidance of pain do not count as true choices. Not so for Hearne. She sees corrections as necessary for establishing the extent of the available freedom and as a guarantee of the contract established with the dog: there have to be (painful) consequences if the rules are broken. From this perspective, it is also not surprising that Hearne (again, like Koehler and Most, but also like the majority of trainers from before the behaviorist turn, including the progressive nineteenth-century hunting dog trainer S.T. Hammond) insists on dividing training into the stage of training proper and proofing. The proofing stage consists of "boundary testing": eliciting conditions that cause the dog to fail and result in applying corrections. Only once the ground rules have been laid down and have become (assumedly, at least) internalized by the dog, can the dog be granted a certain degree of freedom: understood as the human's non-interference in the dog's affairs. In Hearne's view, a dog's house privileges depend on unconditional adherence to a set of ground rules: refraining from peeing on the bedpost, chewing the pillows and jumping up on the sofa. Hearne writes that freedom is being on an "okay" command: you can do as you want unless you infringe upon the rules (54).

Skinner would violently object to the attribution of the feeling of freedom to the dog: an anthropomorphism he would not be likely to forgive. In fact, he would probably view the dog's compliant behavior, established via the use of corrections, as a certain form of apathy or inaction, a by-product of aversive control. Many contemporary reinforcement-based trainers do not wish to settle for an acquiescence achieved through coercive measures. Denise Fenzi writes that if a dog is coerced into a behavior, we cannot speak of a conscious choice (Fenzi 2016, 94). For Hearne, on the other hand, the constant self-management that the operant dog engages is antithetical to freedom. These are simply incompatible visions. In Hearne's world, the possibility of freedom relies on the existence of consequences for breaking the rules. Corrections are inescapable because they establish the boundaries of freedom. The possibility of the dog disobeying is always present: "Salty, too, can always decide to bite me and split. I have known this to happen. She is free, or rather she is not free, in the way babies aren't free" (Hearne 1986, 57). Should Salty decide "to split," Hearne's honest – that is coherent with her worldview – response would be to correct her, thus reminding her of the relationship they have established. She reiterates the opinion first expressed by William Koehler that dogs have a right to the consequences of their actions (Koehler 1976, 57–8). In Skinner's universe, Salty's decision "to split" would not be proof of disobedience, a term that has no place in a behaviorist's vocabulary, but proof that Salty had not been properly trained; that there exist circumstances in which her response has not yet become generalized. Skinner's response would be to arrange the environment in a way that makes the dog succeed and then incrementally raise the criteria in a way that ensures the dog's success.

Hearne's resistance to this proposition would rest on the claim of what is natural and what is not, rooted in an understanding of nature that seems compatible with the early twentieth century but not with Haraway's "naturecultures." Hearne's worldview rests on the notion of natural consequences: punishment is natural. It is an element of nature and of God's grand design. Managing the dog's environment and manipulating the dog's desires are not. For Hearne, it is natural that the dog wants to chase a squirrel. It is not natural that a dog would actively prefer to be with his human. Resisting the temptation to chase a squirrel should, in Hearne's world, result from the dog's understanding of the consequences of misbehavior. Anything other would be unnatural; Hearne writes of the "cunningly cruel arrangements meted out by 'Mother Nature' " (Hearne 1991, 61). Hearne also sees training as instilling a sense of self-control in the dog but doing it "naturally." She senses dishonesty in the relationship the positive trainer creates with the dog: the dishonesty results from depriving the dog of the natural consequences of his actions. It can also be read as misrepresenting the world to the dog by creating an ultimately false language that obscures the cruel reality of "nature." Either way, for Hearne this borders on abuse. She approvingly recalls the words of a famous sheepdog trainer who chastised his students when they praised their dogs: "Jack Knox, a sheepdog trainer originally from Scotland, will shake his crook at a novice handler who makes a promiscuous move to praise a dog, and will call out in his Scottish accent, 'Eh! Eh! Get back, get BACK! Ye'll no be abusin' the dogs like that in my clinic' " (Hearne 1991, 63). Working for food (or even praise!) is something Hearne sees, like the sheepdog trainer Jack Knox and like the Koehlers, as detrimental to the dog's dignity. Skinner, again, was acutely conscious of potential reactions like Hearne's: he argued that we tend to give more credit to an individual for his behavior when the environmental apparatus controlling his behavior is not evident. Offering a reward, particularly a food reward, is like exposing the scaffolding that humans wish to hide: a dirty little secret. To use dog training examples, a dog who sits because he could otherwise be corrected for non-compliance (though it does matter that he is not; the threat is there but it is not carried out) is given more credit than a dog who sits and is then rewarded with a treat. In *Beyond Freedom and Dignity*, Skinner writes:

> What we may call the literature of dignity is concerned with preserving due credit. It may oppose advances in technology, including a technology of behaviour, because they destroy chances to be admired and a basic analysis because it offers an alternative explanation of behaviour for which the individual himself has previously been given credit. (Skinner 1971, 62)

Skinner was cognizant of the fact that the popular understanding of freedom relies on the existence of punitive consequences and aware that arguments about the "unnaturalness" of positive reinforcement would be meted out against him. He saw himself, however, as a social reformer, invested in the notion of progress. Ha was also a scientist and thus felt obliged to analyze the incoherencies of the definitions of these fundamental categories. He was particularly interested in revealing how the notions of moral struggle, dignity and responsibility are operative only in a punitive context and how they would have to be redefined if the aversives meted out by the environment were taken out of the equation. In a 1964 filmed interview Skinner comments on how the entire (human) system of religion, law and morality is based on a punitive model: "where the individual survives only by being responsible by working his way up by controlling himself in order to avoid punishment. If you then propose a different system in which people would naturally be good, it will be rejected because there is nothing comparable to moral struggle involved in it" (Skinner 1964). Thus, where Hearne sees moral struggle (the dog thinking: "should I pee on the bedpost even though I know it is wrong?") Skinner sees the extinction of a behavior that has been controlled through aversive measures. However, Hearne's insistence on attributing this moral dimension to the human-animal bond is by no means unusual: one of the reasons why reinforcement-based training emphasized the importance of self-control exercises, as described in the previous chapter, was that they offered something similar to the notion of moral struggle but in non-aversive circumstances. Practicing "doggie zen" was a way for salvaging that notion of moral struggle; of escaping the purely scientific vocabulary that behaviorism tried to impose and that trainers felt uncomfortable with. Hearne, however, does not accept that substitution. For her, it is only within the traditional aversive framework that the human-canine bond can become imbued with the moral aspects she so wants to see in it.

Hearne's quasi-religious and most certainly anthropocentric position can also easily be discerned in her definition of training, which – as she argues – "results in ennoblement, in the development of the animal's character and in the development of both the animal's and the handler's sense of responsibility and honesty" (Hearne 1986, 43). It is difficult not to notice the correlation of Hearne's views with those expressed in the second half of the nineteenth century: training can only be conceived of as ennoblement of the animal from a perspective that privileges the human. What is ennobling, however, is not the imitation by the dog of the human form of character traits; ennoblement is seen is the expression of something like "canine essence" that can only be fully actuated in interactions with humans. This essence is, yet again, rooted in an immutable nature, in which humans benevolently rule over the animal kingdom. Absolute obedience from a dog is seen by Hearne as conferring character and dignity upon the dog because it is both a gesture of acceptance and of confirmation of the nature of the human-canine relationship. In this framework, training is not meant to be fun. In fact, it cannot be fun: it is an expression of what seems to be the age-old, immutable relationship between human and animal. Bearing this in mind, it is easy to understand why Hearne scorns training through play. She speaks of using play in training as a "sort of activity [which] is part of a way of hanging out with dogs that is fun and that provides both dog and owner with a lot of healthy outdoor exercise. There is no harm in it, but it isn't dog training" (90).

Hearne is also *a priori* skeptical of the language of science: specifically, skeptical of behaviorism's suitability for describing the human-canine connection. Hearne's skepticism has usually been described in a different context: it is most often brought up in relation to her approach to human/animal intersubjectivity. A position she inherited from Stanley Cavell, Hearne posits that consciousness of the fact that humans will never be able to experience the world from the animal's point of view; to know how the dog or horse sees the world, makes us question the nature of our knowledge of the world. Kari Weil sees this as a potent analytical position, noticing that "Hearne's writing on training illuminates the problem of skepticism that has been central to the linguistic turn and takes the discussion of this problem a step farther. For her, training necessitates skepticism regarding our knowledge of the other and, through this, our knowledge of the world" (Weil 2012, 11). However, Hearne's skepticism has even greater reverberations: in assuming that we can never truly know the non-human other, she rejects all scientific claims to knowledge, speaking of intersubjectivity as not based on shared mental phenomena but on the "flow of intention, meaning, believing" and the "varied flexions of looped thoughts." There is a circularity to Hearne's "looped thoughts": because she believes that science has nothing to offer, she refuses to become engaged – or even acquainted – with science. Finally, what is most worrisome in Hearne's approach is how her skepticism leads her, at times, to epistemological solipsism and how that translates onto practical aspects of interacting with animals (and sometimes humans). In Hearne's eyes, the dogs that are trained via a behaviorist approach are deprived of their dignity and intelligence:

> To the extent that the behaviorist manages to deny any belief in a dog's potential for believing, intending, or meaning, there will be no flow of intention, meaning, believing, hoping going on. The dog may try to respond to the behaviorist but the behaviorist will not respond to the dog's response. ... The behaviorist's dog will not only seem stupid. She will be stupid. (Hearne 1986, 58)

This excerpt has been read approvingly as something of a foreshadowing of Barad's theory of subjectivity as produced through intra-action (Vint 2008); using an example particularly potent for the posthuman turn because it involves human-animal intra-action. The subject does not precede the experience but is constituted through and in the encounter with the other. Yet it seems an over-optimistic reading because what the quote above reveals is Hearne's incorrigible anthropocentrism and, contrary to Vint's reading, her lack of openness towards a way of thinking that is not necessarily her own: a refusal to honestly analyze the encounter simply because it does not seem to confirm her presuppositions of the "nature" of behaviorist thought. It is not the behaviorist who claims his dog is stupid. It is Hearne. I have repeatedly mentioned that crossover trainers are almost unanimously amazed with how positive training makes is possible for them to see their dogs as smart. The hypothesis that Hearne's version of skepticism could become particularly troubling becomes glaringly blatant if the actors of this intra-action were two humans of different races. It so happens that in *Adam's Task* Hearne provides us with an example that fits these criteria precisely. She speaks of her recent move from California to New York: "My husband and I, recently translated from California to the Northeast, were given (…) advice about deciding whether or not to walk through Spanish Harlem. If you think it is dangerous, then take a bus, because if you think it's dangerous, then it is dangerous – they can *tell*" (78, emphasis in the original). The assumption that the inhabitants of Spanish Harlem would viciously attack passersby simply in response to the Caucasian pedestrians' fear is a blatantly racist one, even if Hearne tries to ground it in some rough vision of affect theory in which sensory experience constitutes the material world. It also reveals how engaged in boundary-formation Hearne's project is: it depends on an "us versus them" logic. Hearne's discomfort with behaviorism is understandable and absolutely not unique to Hearne: it was an experience shared by both trainers and philosophers (Haraway 2003, 2008). However, Hearne's project, rather than engaging and probing the roots of her own discomfort – as, for example, Haraway's narrative of agility training does – creates an apology for training methods that she chooses to use.

3 David Mccaig: Pastoral Dissent

Donald McCaig began his dog-writing career with a novel titled *Nop's Trials* (1984) that discusses the world of sheepdog training as an aside of the main plotline, which concerns the kidnapping of the title character, a top-level trial dog. McCaig then turned to non-fiction and continued his exploration of the world of working sheepdogs with books largely based on his own experience as a sheepdog trainer. These include three books that could be loosely classified as training diaries: *Eminent Dogs and Dangerous Men* (1991), *A Useful Dog* (2007a) and *Mr. and Mrs. Dog* (2013). The training diary as a genre takes off in the 1990s and 2000s when the audience for such texts becomes solid enough to merit publication. Most training diaries are published by specialized small-circulation presses that deal specifically with dog training – such as Dogwise Publishing – but McCaig's esteem as a writer of fiction made it possible for him to engage mainstream publishers, e.g. Harper Collins. McCaig is also the author of *The Dog Wars* (2007b), a book published with a smaller press for sheepdog trial enthusiasts, which chronicles the history of the border collie in the United States and the conflict between two groups of sheepdog fanciers: those who are interested in conformation shows and those who wish for the breed to remain primarily a working dog. The role of training is a significant one in this conflict. Not only does it matter how the dog is trained, it also matters what the dog is taught and how training shapes the multitude of individuals constituting the border collie breed. In the story of the fight for preserving the working sheepdog, reinforcement-based training methods play the role of the bogeyman.

The fact that McCaig wrote the introductory essay to the 2000 re-edition of Vicki Hearne's *Adam's Task* is not coincidental: he certainly shares many of Hearne's ideas on the human-canine relationship and on training methods. However, as a sheepdog trainer, he brings to the forefront another grounding for the resistance to behaviorist methods expressed in the 1990s and into the 2000s, one that can be dubbed a form of pastoral dissent. Chapter 2 explored how training had traditionally been embedded in what I have called, after Nikolas Rose, eugenic biopolitics; in other words, how trainability has served as a criterion for breeding selection. As the function dog training serves changes towards managing the behavior of entire populations of urban and suburban dogs – which I have discussed as entering the framework of the governmentality of the human population through ensuring humans' safety from dog maulings and canine-related diseases – training slowly ceases to play this eugenic role. As the need for working dogs wanes, conformation shows take over as the primary tool of breeding selection for purebred dogs, although after World War II puppy mills begin to fulfill the growing demand for purebred dogs (Grier 2006, 351). Since the 1970s and the rise of the popularization of spay/neuter campaigns first promoted by the ASPCA (Lane and Zawistowski 2008), pet dogs cease to be expected to reproduce at all. However, it is behaviorism, with its idea that any organism can be taught any behavior that it is physically capable of performing, that seems to radically sever the tie between breeding and training: an idea that does not sit well with the breeders and trainers (often and not coincidentally: breeder-trainers) of working dogs. As Amanda Milliken, one of the main characters of the documentary movie about sheepdogs *Away to Me* and a well-known Canadian herding competitor and breeder, says in the movie: "One of the goals of trialling is to test out breeding stock. You breed one great worker to another great worker and all of those traits that come in the dogs, come off some way in their offspring" (quoted in Hadra 2012). This approach differs starkly from the predominant behaviorist approach in which the effects of training are proof of the trainer's skill, not the animal's inborn talent. Indeed, B.F. Skinner's "science of behavior" was famously disinterested in issues of heritability despite mounting evidence to the contrary that came even from Skinner's former students. The later decades of Skinner's life were marked by his growing interest in social engineering of human society via positive reinforcement, which took his thought away from what in hindsight seems an obvious direction: the issue of species specificity and the importance of ethological observation.

McCaig diagnoses Skinnerian behaviorism as siding with "nurture" in the age-old "nature vs nurture" debate; a position he cannot agree with as a trainer of working dogs: "They [the behaviorists] were anti-Darwinians. In the nature/nurture debate behaviorists came down foursquare for nurture: all behaviors were created and all dog behaviors were equal. Complex behaviors could be shaped from simpler ones, properly reinforced" (McCaig 2013, 73–74). The behaviorist assumption that any organism is capable of learning any behavior counters the shepherd's practical experience with biopolitics. For McCaig it is obvious that it is not possible to teach any and every dog to herd. The dog needs to have "sheep sense" in him in order to herd sheep. Generations of rigorous breeding selection, individual talent and a good handler are all required. The sheepdog is not a clean slate that can be shaped through the skillful use of reinforcement but a carrier of potential resulting from the thousands of years of cooperation with humans and of selective breeding. McCaig diagnoses the trainer's task as the fulfilling of this potential, which – just like Hearne – he sees as resulting in the animal's happiness. As such, enabling the possibility of living up to the dog's genetic potential, is a human's moral obligation.

I have written of the possibility of using Foucault's concept of pastoral power to speak about the human-canine relationship as it develops in reinforcement-based training; pointing to the strong two-dimensional affective bond that arises between the two subjects, the fact that pastoral power is a "fundamentally beneficient power," (Foucault 2007, 173) that is it flows out of the desire to "do right" by the dog; it manifests itself as a feeling of duty and develops an almost overarching sense of responsibility for his flock in the shepherd. Nicole Shukin picks up on the materiality of Foucault's metaphor of the sheep-fold in order to use it to develop her reading of the military K9 as a tool of America's military power (Shukin 2013, 178–179), forming an argument somewhat similar to Aaron Skabelund's about the German shepherd serving as the enforcer of colonial law (Skabelund 2008), except that Shukin's reading adds a biopolitical dimension to Skabelund's postcolonial framework. The analogy between pastoral power and the relationship between a shepherd and his flock is, of course, strikingly direct. It is, in fact, the original example of pastoral power. All of the features of pastoral power that I have described as arising in a human-canine relationship in reinforcement-based training can also be discerned in the type of relationship that McCaig posits as existing between the actual shepherd, his dog(s) and his flock, although the fact that in this case pastoral power is conducted over a multiplicity (which is Foucault's original example) changes the character of this relationship, embedding it firmly within biopolitics. Considering the well-being of the entire flock (and not just of the dog) redefines the relationship between the shepherd and the dog: the dog gains a role in the bigger picture of flock management and becomes a tool for implementing the shepherd's strategy of assuring the well-being of all flock members. While an affective relationship most certainly develops between the human and the canine, it is not a goal in itself but the result of the process of working together. If the dog proves to be incapable of performing the task he is supposed to do – if he is too rough with the sheep, for example – the well-being of the stock in endangered and the logical action in this case is to get rid of the dog if the problem cannot be efficiently resolved through training. The perennial problem with eugenic biopolitics, as explored in chapter 2, is that it operates via a rhetoric of disposability. Once again, the dog that is deemed untrainable, is to be disposed of, while the dog who "shows promise" is selected as a candidate for training. Harsher training methods are also justified, again, via the notion of the well-being of the entire flock.

From this perspective, it is not surprising that the trainers of sheepdogs have been reluctant to abandon more punitive training methods; not only because it is difficult to control the environment of the herding dog in a Skinnerian fashion and to break up the behavior chains into their components, but also because the notion of unconditional obedience is tied up to this type of a power relationship.[4] Foucault writes of the requirement of the "sheep's" unconditional submission in Christian pastoral power:

> the relationship of the sheep to the person who directs it is one of complete subordination, which is again specific to the Christian pastorate and I do not think is found anywhere else. Complete subordination means three things. First, it is a relationship of submission, but not submission to a law or a principle of order, and not even to a reasonable injunction,

4 I write about American sheepdog training methods in detail in the article "Of Dogs and Shepherds: Sheepdog Culture and the American Pastoral" (2015). In a nutshell, the methods of herding dog training largely remain based on the outdated model of the theory of dominance well into the twenty-first century. Herding dog trainers staunchly claim that the behaviorist position is absolutely useless in the endeavor they are engaged it.

> or to some reasoned principles or conclusions. It is a relationship of the submission of one individual to another. The relationship of submission of one individual to another individual, correlating an individual who directs and an individual who is directed, is not only a condition of Christian obedience, it is its very principle. (Foucault 2007, 231)

If one reads Foucault's analysis of the distinction between the Christian pastorate and the Greek pastorate, then it becomes clear that the relationship exemplified in McCaig's writings, and also the human-canine relationship as understood in Vicki Hearne's writings, follows the Christian tradition rather than the way the concept was understood by the Greeks. For the Greeks, just like for reinforcement-based trainers, the point of obedience was self-mastery (233). In the dog training realm: reinforcement-based trainers do not even speak of submission or obedience but rather of the exercise of impulse control: an active process of shaping one's reactions to stimuli. For trainers like McCaig and Hearne, impulse control is simply the result of submission; a passive acquiescence to human leadership.

The problem with this kind of relationship is that it collapses once it is taken outside of the framework of the well-being of the biopolitical collective, which serves as a kind of guarantee of the ethics of this form of power. In the Christian framework, submission makes sense because of its implication in the broader theological project, which includes the existence of a reward in the afterlife. Obviously, in the case of the human-canine relationship, the afterlife does not factor into the grounding of this framework. It would be ludicrous to claim that the dog submits because he expects a reward in heaven. If anything, the relationship is a re-enactment of man's dominion over animals. In the framework of *raison d'etat*, the secularized power that pastoralism mutates into in the sixteenth century per Foucault's reading in *Security, Territory, Population*, the good of the multiplicity demands individual sacrifices. If the dog bites the sheep, *raison d'etat* tells the shepherd he must let go of the dog (unless the fault is deemed correctable; that is unless the dog is trainable). Yet if we take *raison d'etat* (in the form of sheep that have to be taken care of) out of this equation, it collapses. The notion of utility, the need for *A Useful Dog* – the title of McCaig's 2007a book – is crucial in this framework. The acknowledgment of the dog's status as pet, of the primacy of emotional attachment over use value, would make it next to impossible to justify both disposability and harsh training methods. No wonder that the word pet is considered the ultimate insult by McCaig and his fellow sheepdog trainers: "we are indifferent to everything but one preoccupation: we would glorify stock work; and scorn the beautiful and the useless, the lapdog, the untrainable, the barking watchdog, the dumb but lovable family pet" (McCaig 2007b, 143). Pets are outside of the sphere of interest for McCaig, precisely because they are not an element of the framework of eugenic biopolitics. They are "dumb but lovable" even if they have been trained to perform complex behaviors, because it is not the training process per se that is seen as important but its embedding in the broader framework of canine usefulness. Teaching the dog something that cannot be passed on to future generations makes no sense for the sheepdog trainer. In fact, it can be dangerous for the population as a whole.

For McCaig, the rise of behaviorism, a training technology that to him seems absolutely incompatible with the breeding of working dogs, can result in a loss of the abilities that canines developed as a result of breeding selection over a matter of centuries. He writes of AKC border collies, selected through conformation shows, as devoid of the instincts that, to McCaig, make a real sheepdog:

> From time to time an AKC "herding dog" was signed up for a sheepdog clinic. These were perfectly nice dogs, well groomed, well cared for, and their owners loved them. The poor beasts would run around the training ring looking everywhere but at the sheep, or they'd take a pee, or they'd bark, or, sensing that something was expected of them, they'd start doing pet tricks. (McCaig 2007b, 4)

Ethologist Raymond Coppinger speaks of trainability as a characteristic that distinctly separates the dog from the wolf (2001). In this understanding, trainability developed within the context of canine usefulness to humans and through the modification of the wolf's hunting chain via breeding selection. The loss of such trainability and its replacement with what McCaig describes as simply the ability to do "stupid pet tricks," that is behaviors that have nothing to do with the specific species heritage of the dog (and which, in addition, are rooted in the ambivalent history of the circus, dog and pony shows and, more recently, television entertainment via David Letterman's long-running segment "Stupid Pet Tricks") is seen by him as the taking away of the essence of canineness. This position is sparked by an ethical impulse as well; McCaig sees behaviorism as harmful for dogs. Of course, McCaig's understanding of harm is different than Skinner's and has nothing to do with the use of aversion but everything with biopolitics. While McCaig writes convincingly of the relationships he has with individual dogs, his main worry is about the quality of the population of border collies in general. His concern is not so much for the individual dog as for the breed: the multitude.

McCaig is in good company when he writes about loving an entire species, or a breed within a species. In *When Species Meet* Donna Haraway speaks of practices of loving "kinds of dogs" (Haraway 2008, 95) and provides specific examples – profiles of love, one could say – of individuals who have dedicated their lives to such love: searching for genes responsible for a particular disease that affects a breed but also loving through breeding to a particular ideal; working on preserving the dogs' "original" function. McCaig could have easily been profiled in Haraway's chapter for the love he has for a particular kind of dog, the working sheepdog, is clear and overwhelming. Utilizing training as a selective tool is the primary biopolitical strategy for preserving the "kinds of dogs" that both Haraway and McCaig love. There would have been no Cayenne Pepper if it was not for the perseverance of her breeders who worked for decades to produce a dog with a particular temperament; breeders who bred for trainability. Yet when that love is placed in a broader biopolitical framework, it becomes clear that it still has the same side effects that all biopolitical selection does: selection is entangled with the notion of expendability. In order to breed dogs to a certain level of performance, a breeder must breed: produce dogs, maintain some of them as breeding prospects and cull others. Culling, in the contemporary world, does not have to be killing; it can be the careful placement of puppies that do not meet expectations in pet homes, and this is indeed what many respected breeders do. Yet even this more compassionate approach creates the rift between working dog enthusiasts and the rest of the world that is so evident in McCaig's scorn for the family pet. Skinner's utopia, as translated onto the dog world, is to get rid of expendability: if we disentangle the notion of training from the framework of eugenic biopolitics, no dog is going to be considered expendable. If we accept that any organism can be taught anything, we erase disposability once and for all. Reactions such as McCaig's are really an attempt at revealing this approach to be a fantasy and saying that disconnecting training from the broader biopolitical framework is not a solution. It can, in fact, create a set of falsehoods.

The arguments raised by McCaig do reveal some of the shortcomings of radical behaviorism and the consequences they had on the world of dog training. As reinforcement-based training was beginning to make its impact in the dog world, it was doing so via a popularized version of the writings of Skinner (as presented by Karen Pryor in *Don't Shoot the Dog*, 1984), not via the more species-specific observations that came, among other sources, from an essay by Skinner's former students Keller and Marian Breland "The Misbehavior of Organisms" (1961), which focused on how instinctive drift interferes with the learning of behaviors by specific species of animals. Even though the significance of "hardwiring" is already prominent in Jean Donaldson's *The Culture Clash* (1996), it is true that the strictly Skinnerian adage about teaching any organism any behavior persisted within reinforcement-based training much longer than it needed to, often leading to the perception of positive training as intellectually dishonest and ineffective. For example, when reading *Shaping Success: The Education of an Unlikely Champion* (2005), a training memoir by prominent Canadian reinforcement-based agility trainer Susan Garrett, it is difficult not to wonder what is unlikely about her new border collie puppy Buzz becoming a champion agility dog. The narrative Garrett presents is one of a dog with serious self-control issues that needed to be overcome before he could succeed in the sport of agility. Buzz was deemed untrainable precisely because of his impulsiveness and boundless enthusiasm, traits Garrett presented as obstacles to his agility career. Yet the agility world is – and was, even when Buzz was beginning his career in the second half of the 1990s – dominated by fast and driven border collies; a fact that cannot be ignored. Thus, Garrett's claims about how any dog can experience the kind of success her dogs did if only he is trained using the right methods has an obvious ring of falseness to it, as confirmed by her own practice of carefully selecting her next performance prospects. While heartwarming stories of rescue dogs excelling at agility abound, no serious trainer would admit that it is possible to take any and every dog to the highest level.

Reinforcement-based training rooted in radical behaviorism was most vehemently opposed by trainers involved in the breeding of working dogs: sheepdog trainers, hunting dog trainers, sled dog trainers.[5] It was, on the other hand, rather easily accepted by trainers who participated in the new wave of canine activities, described in more detail in the following chapter. These activities include agility, flyball, canine disc, etc.; all of them equally ridiculed by McCaig. These are the activities that developed not in order to test the quality of a breeding program (as field trials and stockdog trials had) but to explicitly foster the development of the human-canine bond. They emerged as disconnected from the framework of eugenic biopolitics and thus more open to the new training technologies. Yet, it seems that the disconnection between breeding and training advocated by the rulebooks of these sports (which, unlike the traditional competitive activities, do not speak of "selecting breeding stock" and emphasize positive training techniques) is not as complete as they propose it is: as evident in the fact that Donna Haraway chose a high-drive working-bred puppy as an agility prospect because she wanted a more competitive success-oriented relationship than the one she had experienced with her rescued mixed breed with whom she began her agility adventure.

5 The ethologist Raymond Coppinger was a breeder and trainer of sled dogs for many years. He explicitly writes about how useless the techniques of behaviorism are for sled-dog trainers.

McCaig's conceptualization of the human-canine relationship is certainly a minority report, although it is also representative of the contemporary culture of sheepdog trial aficionados. It is hard not to see McCaig, and others like him, as consciously resisting the paradigm of technological progress and, in an earlier article, I used another meaning of the word pastoral to emphasize this paradigm (Włodarczyk 2015). The word pastoralism has a history of use within literary studies, where it has been applied since the 1960s to name a certain aesthetic mode, first defined by Leo Marx in *The Machine in the Garden* (1964) and later taken on by numerous other literary critics, as well as expanded and redefined by Marx in a later essay "Pastoralism in America" (1986). For Leo Marx, pastoralism was never a simple idealization of nature: the trope of the machine in the garden was used to signify the tensions related to civilizational progress and technology in the pastoral mode. Marx identifies the American pastoral scene not in Alvan Fisher's painting *Pastoral Landscape* (1854), which shows a pastoral figure in repose, gazing at animals grazing in a landscape unchanged by civilization, but in George Innes's *The Lackawanna Valley* (1855), where a young boy watches a steam locomotive passing through the serene landscape of the valley. The American pastoral scene is thus a scene of reflection upon civilizational progress, available only from the vantage point of pastoral retreat. For Marx, the shepherd is an "efficacious mediator between the realm of organized society and the realm of nature" (Marx 1986, 43), a liminal figure, constantly negotiating between two worlds and belonging to neither. It is promising to think of Donald McCaig as such a figure of pastoral dissent: someone who leaves a successful career in Manhattan to lead the simple life of a shepherd in rural Virginia. His life is a statement of his core values: he does not believe in progress, he opposes the omnipresence of technology; he wants to lead a simpler life more in tune with nature. His views on dog training are also pastoral in their refusal to accept the enlightenment paradigm of change.

CHAPTER 7

Be More Dog: Towards an Affirmative Biopolitics

> It used to be, when I first started in dogs, this was long before *The Culture Clash*, the paragon of a good dog owner was somebody who had their dog quote unquote "under control," that your dog was quote "well-behaved" which meant he was not inconveniencing humans, wasn't moving too much etc. Now, more and more we're putting dogs into everything from MRI scanners and we're trying to discern whether the dog is happy. So the mark now of a good dog owner is somebody who is actually fulfilling the dog's basic needs. So letting the dog be a dog, training with the least invasive ways possible.
>
> INTERVIEW WITH JEAN DONALDSON BY ZAZIE TODD, 2016

•••

> [W]e must make a terminological distinction, which Foucault himself does not employ consistently, between biopower and biopolitics. Biopower, which he first theorizes in the mid-1970s, is a form of power in which the life of populations becomes the central object of rule, not only or even primarily through the destruction but also the creation of life. Foucault explores this deployment of power over life or, rather, through life in the context of sexual behaviours, medical practices, racial discourses, economic paradigms, and so forth. The militancy of the ancient Cynics, however, is clearly an entirely different politics of life. Biopolitics is the realm in which we have the freedom to make another life for ourselves, and through that life transform the world. Biopolitics is thus not only distinct from biopower but also may be the most effective weapon to combat it.
>
> MICHAEL HARDT, "MILITANT LIFE"

∴

I have argued that one of the major developments in the training of pet dogs towards the end of the twentieth century is the disassociation of training

 DOI:10.1163/9789004380295_009

from eugenic biopolitics. This begins to be visible already immediately after World War II, when obedience classes open to mixed breed dogs, but it seems that one of the reasons behaviorism becomes successful in the 1990s is because it views training almost as pure nurture; a notion that is compatible with the changes in the human-canine relationship that take place in the second half of the twentieth century: the focus on cultivating the bond between the individual canine and the individual human. If "nature," inborn canine behavior, appears at all in the reinforcement-based training manuals of the 1990s, it does so in relation to the conditioning protocols required to change it. For example, in *The Culture Clash* Jean Donaldson engages in a lengthy discussion of dogs' "hard-wiring" and proposes non-aversive solutions to potential problems resulting from the dog's "dogness": his propensity to chew objects, greet people with characteristic exuberance, chase moving objects, etc. In the 1990s Donaldson is actually quite unique among reinforcement-based trainers in her interest in ethological explanations of canine behavior. However, in the end, even she admits that a trainer does not necessarily have to understand where a behavior comes from in order to modify it (Donaldson 1996, 17–18). Finally, no positive trainer would have argued for changing inborn canine behavior via "breeding out" the problems. It is the approach of focusing on the individual dog and on the relationship-building aspects that enables the reinforcement-based training protocol. The example of teaching the retrieve that Donaldson provides in *The Culture Clash* is quite telling. She recalls the story of Kurt, a Golden Retriever, who did not retrieve and proceeds to detail the shaping protocol that began with rewarding the dog's "watching with little interest" (29) and finally resulted in the dog bringing back the thrown object "with Border Collie-like intensity, exploding after every throw" (30). In the "traditional" world of hunting dogs, a Golden Retriever with no "natural" interest in retrieving would have been culled from the breeding population by being "petted out," in the best-case scenario, or potentially could have wound up in the pound. Donaldson adds: "What's amazing is not that this dog learned to retrieve and to love it, but that most people would have given up on him after that first session" (30). There is an incredible optimism in reinforcement-based training, an optimism that results specifically from the belief in the individual animal's potential for learning.

However, the disassociation of training from the framework of eugenic biopolitics does not mean that dog training ceases to be entangled in the biopolitical project. On the contrary, the dog becomes addressed as a subject of biopolitics but of a new type of biopolitics: not simply a eugenic biopolitics, but a more affirmative biopolitics, which is focused on improving the quality of life and longevity.[1] If leash laws are the symbolic representation of governmentality, the dog park is a symbol of the turn toward affirmative biopolitics in the US.[2] While governmentality is about safeguarding the general human population and developing a sense of responsibility, a "conduct of conduct" among pet owners, affirmative biopolitics focuses on ensuring the dog's pleasure, which begins to be seen (by dog owners, at least) as something akin to a right. Within the range of these biopolitical practices, training is also increasingly viewed as a strategy for improving the quality of the dog's life, for satisfying the dog's needs, stimulating the dog intellectually, making the dog happier and less bored; for it is boredom that begins to be diagnosed as the major problem of the modern dog's life. While arguments about the dog's quality of life were raised already in the 1940s, the discourse reached a new level in the twenty-first century. For Blanche Saunders, a trained dog was a happier dog; yet the human remained the measure of the dog's happiness. Saunders saw training as creating "a more congenial and a more amusing pet and companion" (Saunders 1946, 12). For contemporary reinforcement-based training, the dog's happiness is the central question, not a corollary of the human's satisfaction with the dog's behavior. The question is not so much about whether training should result in a dog's happiness but about tools for evaluating the dog's level of enjoyment: is a wagging tail and a spring in the dog's step a sufficient indicator? Or should we focus on the levels of oxytocin versus the levels of cortisone? Heart rate? Brain waves? This preoccupation with the dog's happiness is linked to another change: training – as a collaborative human-animal practice – becomes an expression of affirmative biopolitics. It is not just a way of improving the dog's quality of life: it is also the enactment of a certain mode of the human's subjectivation that deliberately attempts to transcend

1 Many projects that aim at developing genetic tests for specific diseases, screening procedures etc., would also fall into the category of biopolitical intervention into canine life. The Dog Genome Project, described by Haraway in *When Species Meet*, is most clearly the epitome of the drive to improve canine life by dissecting it on the level of the genome. The movement to expose the harms of purebred dog breeding (particularly the results of breeding for exaggerated physical traits and of inbreeding) and to end the existence of puppy mills are also part of the same biopolitical project. Within this vast project, dog training is one of the many tools for improving the dog's quality of life; however, a tool that does not operate on the molecular level. The outburst of canine behavioral science research is also an element of this turn.

2 The dog park is, of course, a stand in for canine happiness; its sole purpose being to provide the dog with pleasure and entertainment through off-leash romps with other members of the canine species. However, it should be added that contemporary trainers are also aware that the dog park may not be a source of pleasure for every dog and emphasize that dog parks may actually create behavioral problems in some dogs.

the human-animal boundary, an embrace of humans' animality. If we were to adopt Michael Hardt's distinction between biopower and biopolitics, recalled in the epigraph to this chapter, the place of training shifts from a strategy of repressive biopower to emancipatory biopolitics.

We can also conceptualize the changes in dog training in recent years as a move away from radical behaviorism (Pręgowski 2015) but not through the renunciation of the principles of reinforcement-based training. The change comes from within: as reinforcement-based training becomes more aware of the limitations of radical behaviorism, specifically of the artificiality of defining freedom only as the absence of aversives and the ethical problems related to issues of control, it begins to look beyond Skinner for questions that Skinner's behaviorism does not address. What guides this opening up of behaviorist trainers onto other possibilities is their dog-centeredness: their focus on the dog's needs; their respect for the dog as a significant other and their desire to approach each dog as an individual. All of this leads to an increased pluralization of activities undertaken with one's dog under the sign of training, which is also reflected in a significant transformation of the market of training publications. Choice continues to be the buzzword, but it begins to refer not just to the methods of training but also to the activities defined as training and to the materials one addresses in search of solutions.

The publishing market grows but also becomes more fragmented as smaller presses focusing specifically on dog enthusiasts begin to be established. These include: CleanRun Publications, DogWise, SayYes!, OutRun Press; training related magazines devoted to specific disciplines are also founded and flourish. At the same time, it becomes increasingly more difficult to pinpoint training materials for the new dog owner. It is typical for book publications to target the committed dog owner who already has some experience with dog training and who wants to combat a specific behavioral issue or focus on a particular behavior. Entire books begin to be published on topics such as loose-leash walking, excessive barking, fear of strangers, hyperactivity etc. The growing dog sports community generates a market of its own, with books presenting the various methods that can be used for teaching a particular skill. Additionally, as the twenty-first century progresses, this market becomes increasingly digitalized. The role of web-based discussion groups cannot be underestimated for the development of clicker training but as the market matures, increasingly more digital materials begin to be available for purchase rather than for free (as had been the case with the 1990s internet discussion groups). Training DVDs and downloadable videos probably outnumber training books as I am writing these words (2017), a fact that contributes to the globalization of training methods. Online training courses make it possible to teach group classes to students located all over the developed world; a possibility utilized by most of the world's top trainers today. At the same time, a "seminar circuit" is also thriving with well-known trainers travelling internationally to teach weekend courses. The existence of this flourishing training market reflects the proliferation of commodities in a late-capitalist consumer society. Still, the driving force behind this outburst of training materials is a genuine preoccupation with the dogs' well-being and happiness; the same concern that explicitly motivates the proliferation of gadgets that supposedly make human life easier. The very existence of this market shows how dogs have become incorporated into the machinery of consumer culture; a fact that Haraway writes poignantly about in the "Value Added Dogs and Lively Capital" chapter of *When Species Meet* (2008), aware that her own enjoyment of the sport of canine agility is predicated upon the entanglement of the dog-human dyad in neo-liberal consumer culture.

1 Do More with Your Dog

The emergence of the variety of new canine-human training activities can be seen as an element of what Adrian Franklin describes in *Animals and Modern Cultures* as the "extraordinary further growth in the range of activities associated with animals" (Franklin 1999, 46). These popular new sports activities include: agility (an event based on equestrian showjumping), canine disc (dogs leaping to catch plastic discs thrown by handlers), flyball (dogs racing to retrieve a tennis ball dispensed by a machine, a flyball box) and canine freestyle (dancing routines choreographed to music). These sports were invented in the last decades of the twentieth century but in the world of canine sports this already makes them old. Rally obedience, a less competitive and supposedly "more fun" alternative to obedience, was first proposed in 1999 and adopted by the AKC in 2005. It is in the last decade or so that the number of sports exploded to include treibball (the "herding" of inflatable plastic balls), barn hunt (dogs search a barn for containers with rats: no rats are harmed in the process), dock diving (dogs jump into a pool, longest jump wins), nosework (dogs locate containers with specific scents), frizgility (a combination of disc and agility). In fact, new activities take off each year. In 2016 a titling program was established in Dog Parkour, an event in which dogs navigate obstacles in the urban landscape and in 2017 the AKC launched its Trick Dog certification program. What makes all of these activities related to Konrad Most's police dog training program is that they are also seen as substitutes for the original human-canine cooperative activities of hunting, herding, etc. However, while Most's program was steeped in the framework of utility and based on proving dogs' unwavering usefulness to man even in times of rapid urbanization, these new activities are divorced from this framework. They hail dog owners by responding to two already existing needs: that of providing wholesome intellectual and physical activity for the dog and of strengthening the human-canine emotional bond. The imperatives to enjoy oneself and to develop one's full intellectual potential, so prevalent in contemporary popular culture, begin to refer to the pet dog. Many of these pursuits constitute an alternative outlet for canine drives: more socially acceptable than, for example, hunting; easier to carry out in an urban environment and often allied to developments in technology, either because they rely on technological devices (like a ball-launching machine, plastic disc) or because the training methods utilized are open to new scientific findings, employ behavioral science and electronic devices (use of clickers, electronic contact trainers in agility, electronic timing devices). The most recent noticeable trend in training-related activities is the rise of interest in conditioning and injury-prevention programs among dog sports enthusiasts, which proves the point that training activities are undertaken with the dog's best interests in mind. Finally, even if many of these activities entail competitions, they are not spectator sports. Except for a few championship events, there is hardly ever an outside audience attending dog sport trials. No one sells tickets to agility competitions. The human-canine performance, unlike the "dog and pony shows" discussed in chapter 1 or like the "old" sports of Greyhound racing and dogfighting in the pit, does not function as a form of entertainment for a lay audience.

While some of the more established activities have become quite competitive – agility being an obvious example – the majority of the humans who engage in them do so not in order to win championship titles, but because they believe they are "fun" for their dogs. Studies conducted among serious competitors do suggest that they become drawn into the worlds of particular dogs sports to the extent that they acquire subsequent dogs with the intent of being more competitive, yet they are still preoccupied with their dog's agency and well-being, often willing to switch between activities to find one that the dog really enjoys (Włodarczyk 2016a, though the data referred to in this article relates to Poland).[3] The lifestyle magazine *The Bark* encourages its readers to

3 As an aside it must be noted that the increased competitiveness of the more solidified dog sports (agility and flyball are the best examples) has led to changes in breeding practices and the re-incorporation of training into the paradigm of eugenic biopolitics. Because committed competitors are interested in top performance, they look for puppies which have a higher likelihood of delivering top performance. This is basically the story Donna Haraway tells of

"Try a New Dog Sport This Summer!" (Lou 2017) as if the activities were different flavors of a generally wholesome meal. The first organization to issue "trick dog" titles is named *Do More With Your Dog* and its founder, Kyra Sundance, encourages the teaching of tricks to dogs through the rhetoric of relationship-building:

> This [human-canine] relationship, as with any positive relationship in our lives, requires nurturing to keep it alive and flourishing. Trick training is a way to build upon this relationship, establishing communication methods, trust and mutual respect. It offers a way to bond with your dog as you strive toward common goals and delight in your successes. (Sundance 2007, 10)

Silvia Trkman has published a DVD titled *Tricks for a Better Bond* (2010b) and another one titled *Heeling is Just Another Trick* (2010a), where she argues that if only trainers taught heeling with the same mindset that they have when teaching tricks, they would experience fewer problems with this behavior. While it may seem ironic that the trend in the exercises taught to pet dogs has gone full circle – after all, tricks were routinely taught to house pets in the nineteenth century – the training apparatus is radically different. These may be the same exercises but not only are they taught differently, they are also taught for different reasons, resulting in the need to adjust the interpretive framework. The interest in trick training is a response to the seriousness of obedience exercises and their somewhat artificial embedding in the framework of utility. Modern trainers sense something liberating in teaching a dog to walk on his hind legs or bring a beer from the fridge as opposed to solemnly heeling or retrieving a dumbbell over a jump. Yet this liberational aspect can only be

---

looking for a "high-drive" puppy for agility in *When Species Meet*. In the sport of flyball this has led to the establishment of entire lines of sport-bred mixes: mixed breed dogs selected specifically on the basis of their achievements in a sport, not breed purity. However, unlike in the case of the traditional activities of herding and hunting, it is rare for serious competitors to be breeders at the same time. They usually acquire their puppies (spaced out several years apart) from breeders. Secondly, it is very rare for committed sports competitors to discard dogs if they do not progress in their training as expected or if they cannot be competitive because of their ill health. Again, this is in stark contrast to the treatment of hunting or herding dogs who do not live up to their owners' expectations. This phenomenon is easily explainable via the use value/affective value paradigm. An entire market of dog training books exists that help owners deal with issues related to the dog's motivation for training. A recent book by Denise Fenzi, titled *Train the Dog in Front of You* admonishes owners to stop looking for "that magical dog, work with the one that you have right now; the one right there, looking up at you."

detected in relation to the history of dog training; that is with the background knowledge of decades of teaching "sit" and "heel" using the traditional "jerk and pull" methods. While it was the uselessness of the "trick" behaviors that earned them the infamous label of "stupid pet tricks" – as popularized by *The Late Show with David Letterman* – their separation from the framework of utility has been taken on as something of a badge of honor by those who propose that teaching them is a way of nurturing the human-canine relationship. However, with her video *Heeling is Just Another Trick* Trkman points to the fact that there is nothing specific to classic "trick" behaviors that makes teaching them inherently more "fun" than teaching classic obedience exercises: it is possible to approach competitive heeling with the same attitude as hind-leg walking, thus turning it into a trick.

2 Are We Having Fun Yet?

This preponderance of "fun" as a normative experience raises questions of definition. While "fun" is a bottomless and flexible category that is necessarily subjective, in the context of contemporary dog training it does function in a specific, and at the same time highly paradoxical, manner. It should be recalled that, historically, the word fun had hardly ever been used to refer to dog training. Vicki Hearne wrote of training through play as harmless fun but certainly not dog training (Hearne 1986, 90). The very use of this word suggests a redefinition of training and is related to the relocation of the activity of training from the paradigm of utility to that of affect. In other words, the emergence of "fun" is linked to both changes in training methods and to the bigger change in the functions of training. "Fun" is often described as a necessarily interactional experience: it is usually associated with the presence of others with whom one has a positive relationship. Thus, here it is the relationship-building aspect of training that contributes to its perception as fun, which, again, has everything to do with the subjective character of the experience. The propensity of both humans and canines to play; a propensity that in the case of both species extends past the juvenile stage of development (McConnell 2003, 84–104), is at the root of this explosion of "fun" collaborative activities. Fun is also a quintessentially biopolitical experience. The concept can be read as a vernacular rephrasing of the fundamental biopolitical technology of influencing the mind to elicit desirable behavior; deeply entangled in the consumerist framework. From this perspective, "traditional" training indeed cannot be fun by definition. In a delightful attempt at defining fun in the context of Shakespeare's comedies, which I revert to in a highly interdisciplinary maneuver, Donald

Hedrick writes of fun as "[a]n affect as well as an activity" that is "stigmatized by readily tipping its hat in the direction of commercial culture and leisure" (Hedrick 2012, 17), a fact recognized already by the Frankfurt school, whose representatives were, of course, highly critical of "fun" as incompatible with critical thinking. The experience of having fun is also equated with exercising "consumerist choice" (18), which is not just a prerogative but an obligation of the neoliberal citizen. In *The Plague of Fantasies* Slavoj Žižek speaks of the imperative to enjoy oneself as an everyday patriotic obligation for an American: "Not to be happy today is un-American" (Žižek 1997, 174). Fun's definition as a physical activity and not an abstract thought, as well as something that does not require intellectual complexity, seems to be the key here.

The trend within the move away from radical behaviorism is away from self-control (with its requirement of rational thinking, static contemplation of options) towards fun: less intellectually demanding, highly interactive, exhilarating and focused on movement. It is also seen as a rediscovery of nature, but it is not the nostalgic reading of nature as paradise lost – that permeates writings by Hearne and McCaig – but a more Harawayan, post-natural nature: one which is constructed and not discovered. Dogs have "natural" needs and these needs must be fulfilled: even if inventing new activities is a way of fulfilling these old needs. The story of rally obedience is an interesting case study of the emergence of "fun" as a normative category in dog training. Founded by Charles "Bud" Kramer in 1999, the sport of rally obedience responds to the decline in obedience trial entries by positing a new activity, which is marketed as "more fun" than traditional obedience even though it bears a striking resemblance to the original, and a dog trained in traditional obedience will easily excel in rally. In the "Introduction" to the second edition of the book *Rally-O: An Introduction to the Style of Rally Obedience*, Kramer writes:

> Many of those supportive of Rally have become dissatisfied with traditional Obedience. There may be many reasons for this, some real, some contrived. However, it appears that the primary explanation for the failure of entries in traditional Obedience to keep pace with the growth in other aspects of the dog sport, specifically Agility, is that Obedience is not as much fun. When those new to the sport of dogs are considering the direction their interests might lie, and they see Agility on the Animal Channel and compare it to Obedience…, well, you know the story. I know that there have been articles published that indicate traditional Obedience is holding its own and actually shows minor increases in entries in the advanced classes but not the Novice Class. However, when compared to the magnitude of interest in Agility (have you ever heard of an Obedience

> trial closing with 300 entries within an hour after entries are accepted?), one can only conclude that traditional Obedience has a problem. This problem of a stagnate interest in traditional Obedience is the original stimulus that prodded me to spend that noon hour in December 1998 to develop the basis for the Rally program. (Kramer 2003)

Kramer diagnoses the "death of obedience" as resulting from the sport's perception as stale and static. Measured in behaviors per minute of a routine and in dog's overall ground speed, rally is indeed faster paced than traditional obedience; an MTV version of obedience competitions. Yet it basically consists of the same exercises; or of exercises based on the same set of skills. There seems to be absolutely nothing in traditional obedience exercises that makes them incompatible with reinforcement-based training, a point made clear by Morgan Spector's detailed book *Clicker Training for Obedience* (1999). However, they can be – and have been – taught with aversives. A combination of the long history of obedience training, associated with notions of utility, submission and force-based training juxtaposed with the fast-paced interactivity of agility makes agility the quintessential "positive" sport. It would, indeed, be difficult to achieve the desired training effects in agility with the kind of heavy-handed use of aversives that Saunders or Koehler employed; largely because it is next to impossible to speed a dog up through the use of aversives.

It is thus not surprising that so many of the new sports place an emphasis on speed and the dog's visible (for how can it be measured?) enjoyment of the activity. Dogs catching Frisbees® or diving into pools look like they are having the time of their lives. Giuliana Lund speaks of agility as an activity that "serves no economic or biological function" but "it does serve an accidental moral function due to the fortunate concurrence of ethical and effective training practices in a sport that discourages coercion and requires sustained, enthusiastic collaboration" (Lund 2014, 107). It could be argued that the "fortunate concurrence" is not accidental but structural. The human-canine activities that have emerged in the last decades are meant to serve as primarily entertainment for the dog, only secondarily for the human. Thus, they simply cannot rely on a heavy-handed application of aversives because such methods would clash with their purpose: fun and coerced submission are mutually exclusive. Kramer's brainchild, rally obedience, has also enjoyed success but not to the extent of agility. Of course, choice is the master paradigm, yet agility can best be used to demonstrate the changes that I have been suggesting are characteristic of the most recent years: the desire to follow the dog, which I diagnose as the turn towards affirmative biopolitics.

3 Affirmative Biopolitics

If we think about the history of dog training as metacommentary on the human-animal boundary – one could also say on Agamben's *zoë* and *bios* distinction – the story is simple. Since the 1850s training seemed to offer to a way for the selected animal to transcend the boundary: to become a bit less *zoë* and a bit more *bios*. I have shown how in the second half of the nineteenth century this promise was sham, dishonest; as well as how it was intimately entangled in a colonialist framework. In the late twentieth century behaviorist training methods, transplanted into the dog world as a technology of love, were still focused on proving the dog was capable of human-like rationality, of thinking. Even though – contrary to what Cesar Millan claims – reinforcement-based training is generally not based on anthropomorphizing the dog but on universalizing the learning processes of all organisms; the one location for potential anthropomorphism opened up in the resulting attribution of the capacity for rational, that is governable, behavior, as discussed in chapter 5. In recalling the novice owner's exclamation of "Look how smart my dog is!" Karen Pryor sees a concept of smartness defined by a human scale: the dog is perceived by the owner as smart because he is capable of rational thinking. While many reinforcement-based trainers in the 1990s were highly sensitive to the dog's needs; still, the overall paradigm of what training was – from the perspective of how it performs the human-animal boundary – was not revolutionary. Training was seen as opening up in the dog certain faculties that would otherwise have gone undeveloped; as pulling the dog onto the human side of the binary. As an aside, the individual dog's life also becomes more imbued with *bios* because he matters more to his caretaker, a topic explored in detail in applications of Foucault's biopower to factory farming, where the opposite mechanism is at work: within the frame of biopower the fetishization of the pet takes place at the expense of animals bred for consumption, enabling the functioning of the factory farm system (Stanescu 2013). Still, the blurring of the human-animal boundary enacted by behaviorist training was significant enough to generate a backlash: along comes Cesar Millan and calls for re-establishing a firm human-canine divide; seeing its instability as the root of all problems Americans experience with their dogs.

However, the most recent trends in dog training point to a new and significant shift succinctly summed up in the title of Michał Pręgowski's recent article that discusses the rejection of radical behaviorism in contemporary dog training: "Your Dog is Your Teacher" (Pręgowski 2015). This shift is also evident in Donna Haraway's slogan from *Companion Species Manifesto*: "Run Fast, Bite Hard," which takes canine behavior as a model for human life, as well as in

Haraway's definition of training as a process that is co-constitutive for both subjects: a "becoming with." This is a radical departure from the traditional understanding of what companion dog training is. Historically, training has been about curbing dogs' instinctive behaviors in order to replace them with more "civilized" behaviors that make canine presence acceptable in an anthropocentric world; it has been – as Katherine Grier writes in *Pets in America* – an endeavor of " 'raising' an undeveloped mind into a state of concord with other, cultivated members of the community" (Grier 2006, 74). Now, training is no longer about attempting to pull the dog onto the human side of the animal/human divide by developing some of his faculties. It is about erasing that boundary altogether through the affirmation of *zoë* as constitutive for *bios*. What is discernible in the newer training practices and in the popularity of certain activities is the human's desire to "be more dog": to embrace animality. However, this is not just some generic idea of animality but a very specifically constructed notion of dogness understood as enthusiasm, exuberance, energy, sociability, a positive simplicity, immediate satisfaction of "natural" needs coupled with the non-existence of "artificial" needs.

The title of this chapter is inspired by the highly successful British O2 advertising campaign from 2013, titled *Be More Dog*, which played on species fluidity by presenting the story of a cat who longed to forego his usual feline aloofness and reserve, in order to "be more dog" by engaging in typical canine activities, such as running in the park, playing Frisbee®, whilst exhibiting boundless enthusiasm, energy and a positive attitude to life ("Running? Amazing! Sticks? Amazing!"). The tagline summed up the message: "Maybe we should all be more dog." The success of the campaign – which received a number of awards – is due to a large extent to its skillful portrayal of the positive qualities popularly associated with dogs.[4] This philosophy of valuing canine enthusiasm for the simple pleasures of life also structures many contemporary interactions with dogs, including training practices, which I would like to analyze on the basis of contemporary instructional manuals and videos directed at the pet dog owner and hobby dog sports enthusiast, specifically through a comparison of

4 It is certainly also a result of the skillful juxtaposition of dogness and catness. Not coincidentally, within contemporary pop-cultural discourses the difference between dogs and cats is often constructed through the notion of "fun": something that dogs seem to crave, and cats despise. The popularity of the figure of Grumpy Cat (spawning a series of internet memes, books and dozens of accessories ranging from coffee mugs to notepads) points to the popular recognition of the imperative to enjoy oneself and constitutes a (failed) attempt to resist that discourse. Failed because, of course, it immediately becomes commercialized; coopted by the very logic that it opposes. Grumpy Cat's most popular slogan seems to be: "I had fun once. It was awful."

the training materials by two world-class dog agility competitors and instructors: Canadian Susan Garrett and Silvia Trkman from Slovenia.

In order to do that I first need to turn to the term affirmative biopolitics to show how it ties into the training practices I am about to describe. Affirmative biopolitics is usually associated with the work of Roberto Esposito, although Giorgio Agamben has also been read as potentially positing an affirmative biopolitics through the possibility of the obliteration of the distinction between *bios* and *zoë*, which in Agamben's *Homo Sacer* looms as the only strategy for avoiding the negation that biopolitics constantly enacts, as it fluidly slips into thanatopolitics. However, I do not wish to engage Agamben and Esposito here for it would be difficult to keep this engagement from taking over the analysis of training practices. Foucault should suffice.[5] Following Sergei Prozorov (2015), I believe it is possible to speak of a formulation of an affirmative biopolitics already in the late works of Foucault. Furthermore, Prozorov's reading of Foucault seems to be almost perfectly applicable for describing the changes in contemporary dog training I have been hinting at. Prozorov argues that in Foucault's work on *parrhesia*, most explicitly in the last lecture series from the College de France, *The Courage of Truth*, Foucault formulates an affirmative biopolitics "that does not negate 'mere life' in the name of a privileged form of 'true life' but rather relocates truth to the domain of life itself" (Prozorov 2015, 1). Prozorov works with Hardt's distinction between (repressive) biopower and (emancipatory) biopolitics that serves as the epigraph to this chapter and that I see as helpful for conceptualizing the shift in the roles played by dog training: from eugenic biopolitics (which here should be an example of biopower) to liberatory biopolitics as a mode of subjectivation. I have thus far spoken of how the dog is shaped as a biopolitical subject, that is how within the realm of contemporary training the techniques of biopolitical subject-formation are also addressed to the dog, and also of how training becomes one of the

5 Another obvious question may have to do with the potential use of Deleuze and Guattari's category of "becoming animal" from *A Thousand Plateaus*. This category seems to have been very fruitful within animal studies, spawning a series of "animal becomings," from Elspeth Probyn's "becoming horse" (1996) to Rosi Braidotti's "becoming-insect" from *Metamorphoses* (2002). However, not only do I wish to remain within a basically Foucauldian analysis of training practices as enmeshed within specific power regimes discussed by Foucault (and thinking with Deleuze and Guattari would substantially shift the direction of this analysis), I am personally somewhat uncomfortable with "becoming animal" for the reasons specified by Haraway in *When Species Meet*: what Haraway reads as Deleuze and Guattari's scorn for "the Oedipal animal," the family pet and the fact that concurrent valuation of wildness is explored through the juxtaposition of dog and wolf. It would seem a travesty to apply Deleuze and Guattari to companion dog training.

biopolitical techniques for improving a dog's quality of life. Yet, it is even more important that for the human training begins to serve as a tool for exploring (not appropriating but exploring, with a hint of jealousy and awe) a non-human subjectivity that is perceived as somehow truer, more honest; that serves as something of an aspirational model. As the O2 tagline suggested: "Maybe we should all be more dog."

The figure of Diogenes the Cynic is of key importance for Foucault's argument on *parrhesia,* or truth-telling, as discussed in both *The Courage of Truth* and *Fearless Speech* (2001), the published version of Foucault's 1983 Berkeley lectures. The name Cynic means dog-like and Diogenes the Cynic was known for very literally leading a dog's life: living in a tub, defecating and masturbating in public, defying the human-animal boundary in his search for truth. In *The Courage of Truth* Foucault speaks of ancient Cynicism as "a form of philosophy in which mode of life and truth-telling are directly and immediately linked to each other" (Foucault 2012, 167). Interestingly, the passages devoted to Cynicism contain perhaps the greatest concentration of the words animal and animality in all of Foucault's work. Prozorov argues that the notion of animality is crucial for understanding Foucault's affirmative biopolitics, which Prozorov sees an acceptance of animality and a proscriptive one at that: "animality forms the new model that the human being must emulate to arrive at the true life" (Prozorov 2015, 9). In *The Courage of Truth* Foucault opposes Cynicism to the major thrust of ancient thought, in which animality was always "a point of repulsion for the constitution of man as a rational and human being" (Foucault 2012, 264). In fact, Foucault calls the Cynic's life a scandal precisely because of its affirmation of animality:

> In the Cynics, in accordance with the rigorous and systematic application of the principle of the straight life indexed to nature, animality will play a completely different role. It will be charged with positive value, it will be a model of behavior, a material model in accordance with the idea that the human being must not have as a need what the animal can do without. (265)

The Cynic's emulation of the model is of course absolute, and it is its absoluteness that serves as a guarantee of the truthfulness of the Cynic's words. Diogenes' affirmation of animality consisted of the absolute acceptance of the animal nature of humanity, of the *zoë* of *bios*, which was inseparable from an acceptance of corporeality; with all its drives and natural needs. Through this, Diogenes the Cynic conquers the mind/body duality, which has been of such importance for (not only) ancient definitions of animality. A comment on the relationship between the Stoics and the Cynics is also due: chronologically, Diogenes precedes the Stoics. In other words, Diogenes' affirmation of animality is later transformed in Greek thought into a different understanding of *askesis*, one which includes self-control. This, in turn, paves the way for Christian self-renunciation, although the Cynic model is also important for Christianity in its emphasis on speaking the truth and on poverty, a point Foucault makes in "Lecture 10" in *The Courage of Truth*. In the dog training world, the chronological order of the ancient world is reversed: we move from the Stoics to the Cynics, from an emphasis on self-control to an embrace of animality as a source of truth. However, this return to the Cynics' embrace of animality is, I believe, not unique to dog training. Rather, it underlies much of the current interest the humanities have developed for animals. It is at the roots of the animal turn. It is not surprising that Diogenes has become something of an icon for animal studies. Tracing how this embrace of animality is also discernible in practical human-canine training interactions makes it possible to show how the realms of theory and practice are interconnected. Of course, not all of the precepts of the true life are visible in the type of training I am about to describe but the general thrust of embracing animality as a source of truth is most certainly the leading theoretical framework of the most progressive training methodologies.[6]

4 Garrett, Foucault and Radical Behaviorism

The two agility trainers whose methods I would like to compare to trace the shift I am thinking of, Susan Garrett and Silvia Trkman, are both part of the "reinforcement-based training" camp, by which I mean that they do not include positive punishment (aversives) in their training toolboxes. Yet, Trkman's methods – which are hard to separate from her overall demeanor and interactions with her own dogs: the training apparatus; to recall Peterson's animal apparatus – stand out against Garrett's strict behaviorism. In a way, this could be summed up in terms comprehensible outside the closed circle of dog training aficionados not simply as being more dog-friendly but as "more dog." The agility run is a "subject-transforming dance," as Donna Haraway wrote (Haraway 2008, 176), but it is also a performance of dogness on the part of the human;

6 Michal Pręgowski (2015) uses the term progressive reinforcement training to refer to training methods that are based on reinforcement but that reject the NILIF idea of earning privileges, prioritizing the satisfaction of a dog's needs. He uses a term first mentioned by trainer Emily Larlham in her online published "Progressive Training Manifesto."

forcing the human to engage in an activity that oozes dogness; that stretches the limits of humanity, blurring the human-animal divide through the human's embracing of animality. Yet there are two points that are required to formulate such a reading. Firstly, within dog training animality is not understood generically, but specifically. The O2 commercial posits catness as aloofness and reserve, and these features are seen as the opposite of dogness. Secondly, the qualities associated with different species are historically contingent and context-specific. If we think of dogness as nineteenth-century canine sagacity, it is a very different concept than Konrad Most's understanding of dogness, based in a darker side of animal otherness that needs to be suppressed to harness potential utility. The construct that emerges in the late twentieth century, as visible in the O2 advertisement, defines dogness as exuberance, impulsiveness, enthusiasm, sociability and full engagement in everyday matters.

Susan Garrett's methods have been discussed by Donna Haraway in *The Companion Species Manifesto* and in *Where Species Meet*, where Haraway grudgingly recounts her almost total conversion to behaviorism; a philosophy she used to despise as it applied to the human world. Trainer Susan Garrett's book *Ruff Love* (2002) is instrumental in Haraway's acceptance of behaviorism. Haraway comes into the world of dog training full of ideas that she later deems romantic and describes how her reluctance wanes as she experiences the positive results of behaviorist methods. It is their effectiveness, scientific grounding (in Skinner's theory of operant conditioning) and the resulting joy with which the dog engages in the interaction that make Haraway a convert. "The compensations for the dog are legion" – adds Haraway (Haraway 2003, 44), admitting that the approach she is philosophically somewhat uncomfortable with, at least in the beginning, has positive long-term effects.

The roots of Haraway's initial reluctance to Garrett's methods may need to be explained. By the late twentieth century a strict behaviorist approach to human behavior was being replaced with more nuanced approaches that took into account factors other than just reinforcement and punishment. Cognitive psychology had replaced Skinner's behaviorism. Furthermore, the behaviorist approach in dog training, although it certainly can lead to enthusiastic and fast performance, is based on inducing in the dog the type of self-control that is associated with the control society: it is control through desire rather than through the classic disciplinary technologies. Foucault writes: "power is strong (…) because, as we are beginning to realize, it produces effects at the level of desire – and also at the level of knowledge" (Foucault 1980, 59). The human member of the training dyad remains the inducer or shaper of the dog's desires. While the learning process is based on the dog making choices, and being rewarded for making the correct choice, it is the human who offers the binary choice to the dog. In order for the choice to be a simple binary, the human needs to methodically control the dog's environment through leashes, crates, etc., to take out the other variables and make it impossible for the dog to engage in environmental temptations. In the presence of a squirrel, the dog can either make eye contact and get a cookie for this choice or not make eye contact and whine at the end of his leash. What cannot happen – if it did, it would ruin the training process – is the dog actually chasing the squirrel. The dog cannot fulfill his "natural need." The implementation of Stoic technologies of the self, results in self-control; not conceptualized as the obliteration of the "natural need" but as its mastery by the subject.

In *When Species Meet* Haraway recounts how the training protocol advocated by Garrett helped her dog Cayenne overcome the challenge of learning to stop in the contact zone. The contact zone, an actual element of agility equipment, an area at the bottom of certain obstacles that needs to be touched by the dog with at least one paw (not jumped over), has a metaphoric quality for Haraway through its association with the concept developed by Mary Louise Pratt to define space for the interactions of various cultures. Yet it is also very literal. Cayenne is re-trained to stop in the contact zone through a simple behavioralist procedure: if she does not stop in the zone, she is taken off course and put in her crate. She is not allowed to finish the course. This procedure works only if the dog actually desires to continue running, which Cayenne clearly does.

Haraway's reluctance to embrace behaviorism resulted from her wariness of the type of total control of another being that this philosophy entails. Behaviorism, while claiming to open the road to "all positive" training, does not really erase the human's dominant position. While dominance, particularly as the term has been used in relation to dogs (so-called dominance theory) is not part of the behaviorist vocabulary, the process of training is still dominated by the human's godlike position in relation to the dog's learning process. Still somewhat uncomfortable, Haraway chooses to adopt behaviorism as the primary method of communicating with her dog because of its effectiveness but also because of its omnipresence: the other alternative at the time being the classic Koehlerian punishment-heavy military-derived training, which would have never worked in agility training anyway. However, in the narrative of agility training with Cayenne Pepper, Haraway is the student, the apprentice – not the innovator. In this (unique) situation Haraway comes to accept that which is offered to her, but she does not – so to speak – look outside the box.

Garrett's training apparatus reflects her commitment to changing the dog's behavior through the control of his environment and access to resources. The training aids are numerous and include: crates, head collars, elaborate training diaries. The protocols she proposes are very detailed and the consequences presented for noncompliance are shown as terrifying: if the trainer breaks the reinforcement criteria just once, the entire effort could be ruined. The training techniques Garrett proposes make it possible to manipulate the dog's desires in a way that can make the dog crave that which he would "normally" (that is without training) abhor. This is nowhere more evident than in Garrett's procedure for crate training, presented in a 2007 DVD, titled *Crate Games*. The video, filmed inside Garrett's indoor training facility, opens with almost unbelievable shots of multiple dogs running full speed into their crates. For many trainers teaching a dog to stay in a crate is of paramount importance as the crate works wonders for everyday management of the dog's behavior. Yet for Garrett, it is not enough to teach the dog to just comply with being in a crate, her goal is to make the dog actively desire the crate. She teaches this through a very detailed protocol of classical conditioning that associates the crate with food. Indeed, the results are amazing and it is hard to deny that the dogs want to be crated, but the video is also the pinnacle of what some see as wrong with purely behaviorist training: while such training can certainly create a very intense emotional relationship between guardian and dog, it is also based on highly formalized and structured interactions in which there is no space for spontaneity, unstructured displays of affect and little preoccupation with what the dog "naturally" wants. In fact, that which is natural (chasing, jumping, barking, interactions with other dogs) first needs to be eliminated, in order to be re-introduced only under controlled conditions. The dog needs to learn an almost human type of self-control in order to be allowed to regain the privilege of access to canine behaviors. At the same time, some of the dog's desires are consciously strengthened, while others are downplayed. No force was used, but how "natural" is it for a dog to want to run full speed into a cage?

5 Beyond Behaviorism

As posed in the opening of this chapter, radical behaviorism's hold on agility training is waning as increasingly more trainers restructure their training practices in ways that acknowledge their dog's canine specificity instead of imposing human standards of behavior on the dog. They adapt to the dog, changing their training and handling styles in a way that makes them "more dog." Renouncing the type of control associated with radical behaviorism seems to be one of the leading forms of this turn away from Skinner. In the previously mentioned article "Your Dog is Your Teacher" Michał Pręgowski (2015) analyzes this turn away from control through the juxtaposition of the "Nothing in Life is Free" (NILIF) training program, advocated by the Humane Society of the United States and Kathy Sdao's playful pun on this program called *Plenty in Life is Free*, which is also the title of her 2012 book. Sdao notices that it is not only the use of aversives that can be problematic from an ethical perspective. The NILIF program is quintessentially Skinnerian: while it never advocates positive punishment, it still calls for asking the dog to "earn" all kinds of privileges through good behavior. Access to play, other dogs, and even the guardian's affection is contingent on the dog's performance of simple behaviors. In *Plenty in Life is Free* Kathy Sdao notices the discrepancy between the logic of "earning privileges" and the affection guardians feel for their dogs. On the one hand, behaviorism's insistence on constant monitoring ruins the pleasure of snuggling up to one's dog just because he is there: it ruins spontaneous and unmediated affective encounters. On the other, it can also lead in the dangerous direction of depriving the dog of basic resources. In other words, it can be just as dangerous as punishment. She postulates that engaging dogs in training exchanges in which they must earn fulfillment of their basic needs is unethical. Sdao particularly picks on behaviorism's obsession with constant control of the dog's behavior. Sdao's thoughts succinctly sum up the discomfort that many trainers from the positive training camp have begun to feel with radical behaviorism.

Interestingly, in the world of agility, these are not trainers who have given up on competitive aspects of the sport. On the contrary, one of the pivotal figures in this trend, Slovenian-born Silvia Trkman, also happens to be one of the top competitors in the world, winning multiple World Championship titles with a number of dogs. At the same time, Trkman constantly repeats that such success was never her goal and that people ruin their relationships with their dogs by being too focused on winning. Trkman's methods could best be summed up as: speed, spontaneity, play and throwing human ambition out the door. For Trkman, the appeal of agility lies in its fast pace and the excitement this generates in the dog. Putting emphasis on control over speed – as classic behaviorist techniques do: speed is introduced after control has been established – would go against those very qualities that dogs like about the sport: it allows them to run fast, as Cayenne's behavior in Haraway's contact zone narrative proved. Cayenne stops in order to be allowed to continue to run. According to Trkman, in order not just to do well but to experience the full joy of agility, the handler should embrace those aspects of the sport that

FIGURE 5. Many agility competitors experience the sport as forcing them outside of their human comfort zone, encouraging them to engage the course from the dog's perspective, to run faster, be more agile: to "be more dog."
COURTESY OF MAGDA WIDLAK.

make the dog tick: running full speed, turning the run into a chase, becoming more like a dog him- or herself (Figure 5).

Trkman has not published any printed books but she does have a number of training videos and multiple online articles and interviews, in which she explains, in a very low-key manner, how and why she breaks all the rules of behaviorist training. Her videos are filmed mostly outdoors, against the backdrop of the spectacular Julian Alps. She is often filmed training in very casual situations. Out on walks in the woods, her dogs are seen racing around her, off-leash, usually as a group. When she is working with one dog, it is typical to see the others walking around and engaging in doggy activities: sniffing, playing or just casually resting. In her 2011 DVD *Ready, Steady, Go!*, Trkman presents her philosophy as "work less, play more" and run more. In the opening segment she says: "Agility is just about playing and running and having fun and chasing and ... running, again." She encourages handlers to embrace the spirit of agility and work with their dog's natural instincts rather than against them. By this she means foregoing detailed training protocols, being open to sudden changes. In *Agility Diary*, Trkman describes a particular training situation like this: "Here I wanted to teach a reverse figure eight but she [the dog] had some other ideas, so I couldn't help but reward it because I really like her creativity" (Trkman 2015). Trkman also advocates trusting the dog, respecting the dog's individuality and responding to it in the process of training. While it may sound esoteric, there are very specific cases that Trkman uses to explain what she means. If a dog cannot handle staying at the start line – a stay is recommended by most agility trainers to allow the handler to get into position on the course, but it is also a difficult exercise to teach because the dog really wants to run once they see an agility course – Trkman does not argue for re-training the stay. On the contrary, she suggests that a possible solution is to forego the exercise and just run with the dog because, after all, a stay is not a required part of agility, it simply makes positioning easier for the human (Trkman 2011).

Finally, Trkman's signature is her running contacts, an exercise that best exemplifies her philosophy. The contact zones, as mentioned previously, need to be touched by a dog with at least one paw before the dog descends the obstacle. Haraway described teaching her dog to stop in the contact zone with two paws on the ground and two back legs on the obstacle, a procedure popularized by Susan Garrett and used by the majority of US trainers until very recently. Trkman's DVD *Running Contacts That Make You Smile* (2012) opens with a dedication to her dog La: "an amazing crazy little dog of many virtues, but self-control is certainly not one of them." Trkman continues by explaining that she decided to turn La's weakness (lack of self-control) into an advantage by teaching her to run full speed and touch the yellow zone while running, in full extension. Trkman explains that her major gratification as La's partner is being able to smile as she watches her dog run at full speed alongside her: "We both love to run and hate to stop." Trkman's running contacts have met with much criticism from inside the agility community: other trainers question whether it is possible to methodically teach the behavior; whether her dogs' performances are not just a fluke, a stroke of good luck; and whether the handling of such fast performances is even possible for the average human. It is interesting that this minute element of agility training, the contact zone performance, has raised so much controversy. This discussion reveals that the behavior of stopping can be interpreted as an intrusion of the radical behaviorist discourse of self-control into the sport of agility. I use the word discourse consciously, in its Foucauldian definition, because for almost two decades it seemed "obvious" to the majority of trainers that stopping the dog in the contact zone is the only sensible way to train, even though the rules absolutely do not require a complete stop; on the contrary, they reward the fastest performance.

6 Beyond Agility

Trkman is by no means the only trainer whose ideas exhibit this new philosophy. Within agility training, increasingly more trainers are turning away from radical behaviorism. The "One Mind Dogs" methodology, originating in Finland but rising in popularity among the global agility community, is based on the assumption that incorporating the dog's perspective of the agility course is the key to success. Neither is agility the only activity where this trend can be noticed: it is an example rather than the exception. In fact, the desire to not just understand but even inhabit the canine's mind and body can be traced in numerous training materials or journals related to various training endeavors. Cat Warren, in her memoir of training a cadaver searching dog, writes of moments of desire to get inside the mind and body of her dog in order to know what he knows, to smell that which is unavailable to the human being or to possess canine communication skills: "We watched her [the older dog: Megan], trying to learn from her engagement and disengagement, her covert and canny manipulation of this emotionally stunted puppy. We wanted to know what Megan knew" (Warren 2015, 10) while the pet owner's version of the activity Warren engages in, called K9 Nose Work, stresses in the rulebook that the sport is "all about dogs and celebrating their amazing abilities" (K9 Nosework 2014). As I have argued, many of the contemporary human-canine activities declare their goal as strengthening the human-canine bond and making the guardian more aware of the dog's innate needs, drives and instincts. In so many of them, we can also detect an awe of canine abilities coupled with the human's openness to change in order to allow the dog to flourish. It seems as if much of cutting-edge contemporary training is not so much about making animals behave in ways in which they otherwise would not, as Paul Patton defined training in his Foucauldian reading of the training of dressage horses (Patton 2003, 83–99), but quite the opposite. It increasingly is about making humans engage in activities in which they otherwise would not, challenging them in unexpected ways, both physically and mentally. This is evident in many training memoirs that present the arrival of the dog as a moment of disruption of the human's life and describe its gradual reconfiguration in ways that adapt to the canine's presence. English professor Cat Warren never envisioned herself searching for dead bodies, yet her dog's exceptional sniffing capabilities and high energy made her search for an activity that would both constitute an outlet for Solo's drives and an avenue to make his scenting talent shine. In popular parlance this is what people mean when they say that having a dog makes a person more active. Such common knowledge is also confirmed by empirical studies on the activity levels of canine guardians. Not all become agility competitors or cadaver dog experts, but the general dog guardian population is by far more active than non-dog guardians (Schofield et al. 2005; Cutt et al. 2008). Committed dog owners allow themselves to be pulled by their dogs in directions that they would not otherwise take. They follow.

What makes these practices part of the bigger framework of affirmative biopolitics, if we were to follow Prozorov's reading of Foucault's *The Courage of Truth*, is their acceptance of animality not as humanity's other but as constitutive of who we are. The similarity between humans and canines, which was already hinted at in radical behaviorism's pronouncement that all organisms learn the same way, is redefined as a similarity based not on capacity for "rational" (governable, predictable) behavior (as behaviorism had implied), not on the capacity for suffering (the argument first coined by Jeremy Bentham that was formative for developing the discourse of animal rights) but on the capacity for experiencing simple bodily pleasure. In this, dogs are seen as somehow more perfect than humans because their capacity to enjoy the simple pleasures in life is not hindered by cultural norms and codes. Their senses, instincts and drives are also keener, less dulled by civilization. A very similar argument has been recently made by Jeffrey Nealon (2016), though not about recent changes in dog training but about the animal turn in the academy. Through a careful reading of Foucault's *The Order of Things* and *History of Madness*, the texts that first formulate a rough definition of biopolitics, Nealon traces how Foucault's notion of biopolitics was not based on a categorical rejection of animality but on its incorporation into the definition of life. Similarly to Prozorov, Nealon also reads the figure of Diogenes in *The Courage of Truth* as something of a model of the biopolitical relation to animality. In this relation "animals function less as our ethical 'absolute other' than as our hidden, better self: our unconscious drives are animal in nature and it is those unbridled desires that make us most truly who we are" (Nealon 2016, 11). As a result, Nealon views the rise of animal studies as a return to Foucault's original understanding of biopolitics, in which – as he claims – it is not animal life that is repressed but the potential of plant life:

> [a] Foucauldian provocation might suggest that animals are important within contemporary academic discourse […] not because animals function ethically as "wholly other" to humans but primarily because they're "like us" in an originary way: they experience intense feelings; they are born and they die; they like to go walking in the park at sunset. Indeed, animals are wild, just like we are in our best and freest moments. (Nealon 2016, 11)

Meanwhile, trainer and behaviorist Patricia McConnell, writes in *The Other End of the Leash*: "Our species shares so much with dogs. […] Our babies have lots to learn while growing up; we hunt cooperatively; we play silly games even as adults; we snore; we scratch and blink and yawn on sunny afternoons" (McConnell 2003, xvii). Change the sunset to a sunny afternoon, and we easily slip from critical theory to a dog training manual. While the status of plant life is of secondary importance for the argument I am formulating, the shift towards embracing animality is certainly something that the animal turn in the academy shares with dog training; a similarity that provides yet another reason why an outright rejection of the analysis of training is simply not a productive move for animal studies.

7 Back to Ethology, Back to the Body

One could think of this embracing of "wildness" as something of a return to Nietzsche's fascination with animal morality as simpler and not corrupted by civilization but – contrary to Nietzsche's thought – there is absolutely no glorification of violence in the contemporary account of dogs. Aggression is a problem to be countered through training, and contemporary ethology, which regains its significance for dog trainers, informs us that dogs are generally peace-loving critters with a highly ritualized set of behaviors that help them avoid direct physical confrontation (Coppinger 2001). In other words, it is not dogs' capacity for aggression that constitutes dogness but their capacity for pleasure. I have written of the demise of ethology within radical behaviorist training methods. Not only was ethology deemed unnecessary (if all organisms learn the same, the study of species-specific behavior is not going to be helpful to dog trainers), it was also associated with the long history of coercive methods. Ethologists from Konrad Lorenz to Rudolph Schenkel suggested that wolf packs were highly hierarchical and that conflicts were primarily related to usurpation of dominance in the pack hierarchy and were resolved through fights. These observations had a direct relevance for the training of dogs: if dogs are like wolves, they need to be trained with force because it is the language they truly understand. Consequently, it is not surprising that early positive trainers shunned ethological explanations of canine behavior. However, at the turn of the century ethology changes in several ways. Firstly, the wolf studies that served as the grounding of the "alpha wolf theory" are criticized as based on observations made in captivity, and the belief in a rigid hierarchy in a wolf pack is challenged. More importantly for dogs, with the establishment of Adam Miklósi's *The Family Dog Project* in 1994 ethology begins to look at dogs as dogs and not just as captive wolves. Ethology is primarily interested in the behavior of animals "in the wild," and this fact had for decades deterred it from paying closer attention to dogs as a domesticated species. Miklósi's brilliant move was the identification of the human family as the dog's natural habitat: another example of the redefinition of nature that has taken place even within the classic "natural" sciences. Still, the question remains of what ethology might have to offer to dog training, if not alpha rolls and other displays of dominance. The answer refers very directly to the title of this chapter. Ethology teaches trainers to "be more dog." Modern ethology has become a science that teaches animal owners how assuming the dog's perspective can help resolve certain training issues.

It should be added that not all early behaviorist trainers were skeptical of ethology. Jean Donaldson's *The Culture Clash*, a highly influential book published in 1996, is based on the premise that understanding the dog's "hardwiring" is key for modifying his behavior. In fact, the very title of the book suggests that human-canine interaction is very much like intercultural communication and knowing the interlocutor's language – understood very broadly – is key to avoiding conflict. However, the greatest credit for re-introducing ethology into positive training in the US is due to Patricia McConnell, whose book *The Other End of the Leash* (first edition: 2002) is the perfect model for analyzing how the ethological perspective can be incorporated into training practices and which, at the same time, shows how the subtle shift away from pure Skinner is also a shift toward the dog. McConnell's emphasis is on body language: how minor shifts in the human's posture and movement can be interpreted by the dog and how these minute gestures can have huge consequences for the dog's behavior. I have written about the disenchantment with touch as a technique of training that came about with behaviorism and that stemmed from the long history of "bad touch" in animal training. Even the hand gestures, which were traditionally used as signals in obedience training, were not completely arbitrary: they were remnants of the aversive training techniques used to elicit the behaviors; basically threats of punishment. They reminded the dog of the human's constant control, pushing the dog into docility. Classic behaviorist methods made it possible to train the dog without touching him; not only that, they minimized the use of body language while training according to the logic that if one can teach a down without jerking the dog, not relying on body language makes it easier for the dog to generalize the behavior. It should not come as a surprise that they also emphasized the importance of verbal stimulus control.

In *The Other End of the Leash* McConnell uses ethological insight to argue that using body language (and not vocalizations) as a tool of communication is simply natural for dogs. Eliminating body language from the human's toolbox impedes human-canine communication. McConnell cites her own study in which puppies were taught to sit using hand signals and a verbal command to prove that it is simply easier to confer our intention to the dog using our bodies (McConnell 2003, 5–6). In this, she is advising the dog owner to "be more dog": that is to assume the method of communication that is most easily decipherable by the dog. She speaks of the problems that humans have with abandoning our culturally coded yet deeply internalized gestures and motions: of transcending humanity, so to speak. Yet she sees this as an ethical imperative for those who wish to cultivate positive relationships with dogs. Of the many examples provided by McConnell, the case of human-canine greetings seems the most representative of "the culture clash." McConnell writes how greeting another being with an outstretched hand – an act she calls hardwired but which could, in reality, be simply deeply ingrained within specific cultures – is "so strong that it can override clear signals that tell us to stop" (xxiv). On the other hand, such greetings are considered an act of rudeness within canine culture and can thus lead to problems (read: dog bites). McConnell writes that good trainers are good "because they are aware of their own behavior. They've learned to stop doing some of the things that are natural to our species but are misinterpreted by dogs" (xxv). Working on "being more dog" is thus an ethical imperative for a dog trainer: it is necessary for preventing potentially lethal misunderstandings. Changing the human's behavior is a prerequisite for changing the dog's behavior but in a completely different way than it is for Cesar Millan, who argues that the human needs to assume the alpha position in order to firmly enforce the human/canine boundary. McConnell calls for the opposite: the blurring of the boundary through the human's willful and often difficult transcendence of humanity is necessary if we want to achieve peaceful cooperation.

The diversity of McConnell's solutions set against Cesar Millan's alpha posturing reveals how different these two approaches are, even if both claim to base human-canine interaction on assuming the dog's perspective. Millan's solution is a cookie-cutter one: he mistakenly assumes that the essence of dogness is pack hierarchy and imposes this lens on all behavior problems, to a certain extent simply because it makes for flashy television. I have argued that one aspect of Millan's popularity specifically has to do with the ease of his solutions: rooted in a mythical understanding of dogs as wolves in collars, they do not require an understanding of the specificity of the individual dog. However, the harm that could potentially be inflicted by Millan becomes visible when a "whispering" solution is juxtaposed against a solution that actually relies on thorough behavioral analysis. McConnell writes of a client's dog who "seemed to attack visitors rarely and randomly" (76) and about her own quest to figure out a pattern to the dog's attacks. After a lot of hard work on eliminating all the usual suspects, the conclusion McConnell reached was that the common factor was the smell of pizza: "It seems a pizza delivery boy had kicked her when she was about six months old." In the end, the dog was conditioned to associate the smell of pizza with a positive experience and, voila, "she's been fine ever since" (76) – writes McConnell. The solution was simple, straightforward and – most importantly – effective. Yet the detective work that went into arriving at the solution included a detailed analysis of the individual dog's life experiences. If we imagine Millan's probable procedure for dealing with this dog, it would most likely involve a staredown on all fours and some heavy jerking in the presence of visitors, coupled with prescriptions for "calm, assertive energy" and a diagnosis of "imbalance" in the dog's owner. The procedure would have most probably worked in making the dog stop lunging: not because of the calm-assertive energy but because the dog would have been scared into learned helplessness: better do nothing if doing something leads to pain. While the issue of attacking visitors would have been solved, the dog's unusual fear of the smell of pizza would have remained. In McConnell's protocol, reaching the solution required not only an investigation into the dog's past, but also a suspension of the belief in the apparent irrationality of the behavior. Skinner is constantly on her shoulder, so to speak: the dog is an organism influenced by the environment and reacting predictably to these influences. Assuming the dog's perspective is not equivalent to kneeling down on all fours and staring the dog down in a display of dominance. It is not a generic staredown. It is a painstaking analysis of the individual dog's formative life experiences coupled with a belief that such formative experiences can, in fact, be the province of a dog's life. It is not that McConnell would resent going on all fours. On the contrary, she most certainly does engage in such behavior but in a different context that Millan:

> Over the years I've occasionally succumbed to curiosity and got down on my hands and knees and sniffed where my pack of dogs were sucking air like vacuum cleaners. A few times I wasn't able to smell a thing, but most often I'm impressed with some rich, thick scent that smells markedly different than the ground a few inches away. My dogs appear to enjoy my behavior more than I do, increasing their sniffing and tail wagging, and licking one another and me as if something notable had happened. Perhaps it had. (69–70)

Getting down on one's hands and knees is clearly not part of an attempt at enforcing dominance here. It results from an obvious curiosity about the dogs' behavior; the desire to test if a human is capable of replicating the dog's experience, coupled with blissful disregard for human etiquette. It is also, to use a term usually reserved for speaking of literature, an act of defamiliarization: most obviously, of the human's behavior for the dogs, as evident in their somewhat baffled and incredulous reaction, but also of the human's blasé and habitual perception of the dog's sniffing as just something that dogs do. Taking a closer, longer, look – or sniff, in this case – reveals that there indeed exists a reason for the dogs' behavior and that this reason is not beyond the perception of a human, as we casually assume. Getting down on all fours makes it possible to amplify the smell to the extent that it is perceptible for the human nose. It makes the smell more smelly, to play on Shklovsky's definition of defamiliarization as making "the stony stony." In *Pets* Erica Fudge uses the reference to the literary technique of defamiliarization to describe the power of literature about animals (Fudge 2008, 68–69), but the kind of training approach proposed by, e.g., McConnell truly has a similar effect: it makes the human more attentive to the minute canine behaviors that are usually taken for granted, glossed over as dogs doing dog stuff.

8 Conclusion

While it is possible to easily criticize the analogy that I propose between the radicalism of Diogenes the Cynic and the contemporary dog trainer by saying that the dog trainer, unlike Diogenes, remains very much a part of "the system" (in this case a neo-liberal individual: enterprising, self-managing and certainly not on the fringes of the *polis*). Yes, this is the case. Dog trainers are not (and most likely will never be) at the forefront of a revolution. Any revolution, really. They have dogs to take care of. But they are not apolitical either. Good dog trainers are committed to the practice of truth-telling, *parrhesia*, a practice that – as Foucault tells us – is necessarily political. The truth-telling relates, of course, to dogs and the commitment requires an openness to change, a flexibility of thought, coupled with a genuine desire to "know about the dogs," to quote Haraway's famous line from her 2000 lecture at the European Graduate School (Cassidy 2006, 324). Good dog trainers are, of course, interested in the truth about dogs. They err and blunder, and their truths are subject to the same conditions of knowledge production as other discourses, but they try. Because they are committed to the truth, they often shine a critical lens on mainstream discourses about dogs. For example, I have conceptualized the emergence of reinforcement-based training as related to Nicole Shukin's notion of "feeling power." The desire to train "with kindness" is most certainly rooted in the emergence of "feeling power." However, reinforcement-based training is by no means an application of "feeling power": it is necessarily its deconstruction, a shattering of the illusions created by the fantasy of boundless canine love. Skinner's behaviorism as a technology of animal learning provided a reality check as it provided a methodology of training. Nothing shatters illusions of canine bravery and loyalty like knowing the mechanics of teaching specific behaviors. It is the dirty practicality of working on common goals together with other species that destroys fantasies. In the article on "feeling power" Shukin writes of military dogs as the emblems of the fantasy of feeling power. The public views them as capable of unconditional love and of sacrifice. I am certain the trainers do not. The trainers of mine detection dogs know very well that the dogs search for mines not because they want to protect mankind but because they want a ball. And they do get a ball if they locate a mine. Cat Warren writes of how she is often asked if Solo, her cadaver dog, gets depressed because of the type of work he does. She answers:

> No. Solo's work – and his fun – begins with someone's ending. Nothing makes him happier than a romp in a swamp looking for someone who has been missing for a while. For him, human death is a big game. To win, all he has to do is smell it, get as close as he can to it, tell me about it and then get his reward: playing tug-of-war with a rope toy. (Warren 2015, xi)

The recent turn away from radical behaviorism is really a continuation of this quest for telling the truth about dogs. Contemporary progressive training does not reject Skinner's observations on the fallout of aversives. It still defines the infliction of pain as an irreducible. While adding emphasis to the dog's specificity as a dog, and even while redefining training as learning from the dog, contemporary trainers do not become enthralled with a fantasy representation of animal alterity that seems to be perpetuated with a vengeance in media representations of animal morality, bravery and interspecies bonding. The trainers continue to "keep it real." It is not coincidental that McConnell writes: "If you want to get a good belly laugh out of a group of dog trainers, start talking about 'unconditional positive regard from dogs' " (McConnell 2003, 7). It seems that this truth-telling has recently required a giving up of the defining object of the discourse: a radical maneuver but one academics know well from the history of critical theory. Dog trainers have recently acknowledged that to interact with dogs better, they need to give up the rigidity of species distinctions and to reverse the definition of what training is: from making the dog conform to an anthropocentric world to destabilizing this anthropocentrism. As the examples of training practices by Trkman and McConnell show, this is seen as necessary for effective training. However, the measures of effectiveness have most certainly changed.

Conclusion: The Death of Obedience

As I conclude this huge project that has sometimes taken me in unexpected directions, I am left with a sense of incompleteness: it has not been possible to do justice to all the twists and turns in the genealogy of dog training, even in a book project. I find myself holding my breath and saving ideas for side projects, hoping to be able to develop certain connections that I may have hinted at but could not elaborate upon for reasons related to the structure and scope of this book: fitting almost 170 years of dog training into one volume is an impossible task. Because of the methodology I have adopted as a scholar of (mostly) published texts, the story I have created is a story of the written word, which necessarily focuses on those who have been able to participate in the popular publishing market. In the end, this is a story of urban, middle-class and predominantly white America. While it is true that pet-keeping and the training of companion dogs are related to the rise of the American bourgeoisie, this is not to say that "the other America" has not kept and trained pet dogs. However, this alternate story needs to be approached with a different methodology than the one I have employed: it is a project that remains to be written and carried out by another scholar. I also feel sad for not having been able to do justice to the amazing diversity of contemporary training voices. Because the framework of this book is based on detecting patterns, I necessarily had to choose primary texts that make it easier to trace these patterns, downplaying the role of others. In other words, the abundance of materials available since the 1980s has made it impossible to refer to them all without disrupting the proportions of this book. Still, I have not tried to form patterns where there are none. I have most certainly tried to maintain an academic honesty throughout the book and have not consciously left texts out simply because they do not fit the paradigm I am trying to sketch. This is a project of Foucauldian truth-telling, after all. If anything, the reasons I may have omitted certain texts is because they are better suited to being analyzed with non-Foucauldian tools and for reasons of clarity and brevity I have tried to rely on Foucault's work as the scaffolding holding this project together. That said, I truly believe using Foucault makes sense: this is not a contrived attempt of trying to enforce a unity over something that would best be interpreted via a multitude of strategies. On the contrary, Foucault's account of the rise and fall of discipline helps understand the changes that would otherwise remain unintelligible.

 | DOI:10.1163/9789004380295_010

One obvious statement about contemporary companion dog training is that it bears witness to the death of obedience: not as a recreational activity but as a conceptual category for describing human-canine relations. Conceptualizing interactions with dogs in terms of eliciting "unconditional obedience" is a concept of the past and the influence of behaviorism on dog training can be assigned much of the credit for this change, although this turn is also embedded in a series of broader transitions that I have tried to describe. When a dog responds to a command uttered by a human, no reasonable contemporary trainer would interpret the interaction in terms of obedience to a human's will or the dog's natural submission. Most would speak of generalization of a cue; history of reinforcement, etc. Even Cesar Millan's wishy washy "energy transfer" is a way of circumventing dog owners' wariness of the categories of obedience and disobedience, even if it is also a form of backlash against behaviorist methods. However, what I hope this book has managed to achieve is to show how obedience itself is not a timeless, eternal category that had accompanied dog training "since the beginning of time." It rose to prominence only in the twentieth century and it died by the time the century ended.

If I were to speculate on where dog training may go from here, there are several directions I could potentially suggest. Firstly, increased "scientification." I have argued that the belief that dog training and science come together in the late twentieth century is a myth. Dog training has always reflected, in a popularized version of course, the state of scientific knowledge. Footnotes are rare in training manuals but it is clear that Darwin, Thorndike, Lorenz, Skinner, Bekoff, DeWaal, Miklosi (in that order) have been leaving their mark on training practices. It is the science that has changed and will continue to change, shaping the popular understanding of scientific discourse reflected in training practices. Having been shaped both by Haraway's notion of "situated knowledges" and Harriet Ritvo's *Animal Estate*, I do not treat science as a transparent truth and view it as a discourse involved in the bigger ideological projects of the period. Barely twenty years ago, science "told us" that dogs are amoral and self-interested, and that they do not learn by imitation: Jean Donaldson recalls these as "scientific facts" in *The Culture Clash* (1996). Now, increasingly more scientific studies suggest that it is otherwise, and dog training is heeding the call. Dog trainer and scientist Claudia Fugazza has already published a book, titled *Do As I Do* (2014), which advocates "social learning" as the optimal method for teaching behaviors. The specificity of canine play also seems to be promising in terms of its potential applications in training techniques. Even Millan, who bases his career on the anti-scientific stance, has recently begun to incorporate more canine science into his newer books.

Wherever "science" will go now, dog training will follow. This collaboration between contemporary scientists and trainers is particularly interesting because of its intensity and the scientists' awareness of writing for an audience of dog enthusiasts. Even if dog trainers had read Thorndike and Skinner, it is clear that Thorndike and Skinner were not addressing dog trainers. Contemporary canine scientists are. There even exist numerous books and training blogs that translate the newest scientific research into plain English for the average dog trainer. Some scientists moonlight as dog trainers or turn to training only after getting a PhD, which would have been considered an unusual career choice several decades ago. Thus, predicting developments in dog training is really contingent upon predicting developments in canine science. The current trend certainly seems to be toward the blurring of the human-animal divide through the discovery among canines of subsequent traits that we had hitherto assumed were unique to humans. The recent discoveries have focused on dogs' capacity for affects, including a sense of fairness and rudimentary morality. The most spectacular discoveries have been related to what is presented as a canine capacity for love (Berns 2013). As a result, there is a tendency that dog guardians are beginning to adopt from the range of scholars working on animal behavior and cognition – and particularly from popular media representations of such research – to think of dogs as increasingly smarter. The dog has gone from "a lemon brain" – to recall a term employed in 1996 by Jean Donaldson – to an intellectually and emotionally complex being with abilities akin to those of real superheroes. It is a sign of the times that Brian Hare, a well-known canine scientist, published a book titled *The Genius of Dogs: How Dogs Are Smarter Than You Think* (2013). At the current moment, there exists an almost symbiotic relationship between scientists, who are producing increasingly more studies about dogs' exceptional abilities; popular media, who revel in stories of canine heroism and intelligence; and dog owners, who desperately want to believe that there is more to the animal they share their lives with than meets the eye. This belief in canine genius is an element of what I have referred to as the fascination with animality that seems to be the guiding principle of so much contemporary thought and practice and which I actually see as linking "the animal turn" in the humanities with the practices of progressive dog training. Interestingly, the role that dog trainers play in this web of complex relationships is to "keep it real" through constant curbing of excitement generated by the steady stream of popular media accounts ("viral" videos on social media, etc.) of feats of canine intellect and emotion. Nothing shatters illusions like knowing the mechanics that go into teaching specific behaviors. Of course, this is a generalization. There are so many trainers and so many methods out there that one can find plentiful counter-examples. After all, dogs are notoriously easy to train – it is what makes them dogs – and, historically, they have been successfully trained with the strangest methods that had no right to work and yet did. Still, the fact that trainers are immersed in Foucauldian affirmative biopolitics is what makes them (or at least, the best ones: the honest ones) committed to truth-telling. They want to "know about the dogs." Bob Bailey used to remind trainers that "Pavlov is always siting on your shoulder," by which he meant that they should never ignore the dog's history of environmental classical conditioning while engaging in operant conditioning procedures. Today I also see Diogenes sitting on the best trainers' shoulders: they never forget that our relationships with dogs are based on an admiration of their "dogness," even as they acknowledge the need to modify that "dogness" to make our co-existence more feasible.

Secondly, stratification and pluralization of "positive training": I have analyzed how the training of pet dogs gradually exited the framework of eugenic biopolitics; how training and breeding ceased to go together in the case of companion dogs. The current explosion in canine sports activities, while rooted in the contemporary focus on the individual dog and on relationship-building, seems to reverse that trend. As the new training activities become solidified, a new class of competitors is developing within the activities that were meant to serve as harmless "fun" for the dog. The top competitors are beginning to be able to devote themselves professionally to dog training, ceasing to approach training as a hobby and beginning to view it as their livelihood. While the majority of participants in sport activities do so with their family pet (often obtained from a rescue organization and not a breeder), there are increasingly more exceptions. As sport enthusiasts become more involved in their discipline of choice, they begin to look for dogs that could be more competitive and they turn to breeders who try to cater to that need. Most serious agility competitors can list entire lines of dogs that excel on the course and breeders of mixed breed sports dogs cater to flyball enthusiasts. In other words, the activities that were invented as artificially constructed outlets for a dog's natural needs are becoming re-incorporated into the biopolitical framework: they are starting to serve as tools of breeding selection. The breeding of classic working dogs has also changed. As sheep-raising is losing its significance for the American economy, sheepdog trials continue to serve as a tool for breeding selection. However, the competitors are often people for whom the relationship with their dog is prior to their involvement with the sport: they are converts to McCaig's gospel of canine utility (Włodarczyk 2015). These ethical conundrums cannot be resolved here but it is worth noting that they exist. At the same time, the potential contemporary training protocols have for being modified to suit individual dogs with individual histories truly extends the notion of trainability in a way I feel I have not given an in-depth account of in this book. Within the logic of eugenic biopolitics and the framework of utility, trainability is not only a breeding criterion but also a matter of life and death. Chapter 2 explored this on the example of nineteenth-century hunting dogs, but this paradigm can still be found in (some) animal shelters: if a dog is deemed untrainable, he may find himself on "death row." In these cases, trainability is often equated with lack of aggression. Yet, there exists an entire genre of narratives written by owners of dogs deemed "unsavable," who recount how carefully designed training protocols helped modify the dog's behavior to the point where he could be relatively successfully re-integrated into society. *Wallace* (2012) a *New York Times* bestseller by Jim Gorant tells the story of a "death row" shelter pit bull who went on to become a disc dog champion. An individually-tailored behavior modification program coupled with unwavering commitment of the dog's guardians made it possible for Wallace to overcome the behavioral issues (aggression) that caused him to initially be labelled untrainable. While the somewhat sappy and naïve optimism of this "heartwarming" tale sounds Disneyesque, it captures the optimism of reinforcement-based dog training: its emphasis on the potential of each and every dog; its refusal to give up on a dog. This optimism is often tapered off in stories that end not with ribbons and trophies but with an acknowledgment of the limitations of training, while reiterating the writer's commitment to the relationship with the individual dog. This is the message of Debbie Jacobs's *A Guide to Living with & Training a Fearful Dog* (2008), which – mixing memoir and training manual, in a way that has by now become almost commonplace in dog training literature – tells the story of a dog the author adopted from Louisiana after Hurricane Katrina.

Thirdly, the disappearance of the word training. In the "Introduction" I said that the word training is a signifier with constantly evolving signifieds. Still, it seems that many people who work with dogs have noticed the limits of the flexibility of training as a conceptual category. Because training has long been associated with physically coercive techniques and with the teaching of specific behaviors (heeling, etc.), even dog trainers themselves are becoming wary of the word in an effort to disassociate themselves from what they perceive to be a rather infamous legacy. In a more theoretical language, training is a technique inexorably tied to the disciplinary regime. As discipline wanes, training redefines itself in the new context and one element of this redefinition is a change of vocabulary. This does not negate the possible directions mentioned above: these may simply take place under a different label than training. Many trainers speak of behavior modification, rehabilitation, socialization, or canine cognition and learning, in an effort to separate their work from the legacy of "traditional" training. However, others – like Patricia McConnell – stand by the word training, recognizing their own entanglement in this tradition and questioning whether a linguistic maneuver has the potential of resolving all ethical conundrums.

Replacing use value with affective value as the defining feature of human-canine relationships has an obvious side effect for the concept of training: if dogs are valued not because of how helpful they can be in carrying out specific tasks but because of the potential emotional connection that they can provide, technically, they do not really need to be trained in order to serve the affective purpose. However, it seems that the ones that are serve this purpose better: the strength of the emotional connection increases almost proportionately to the amount of training a pet dog has received. Historically, the training of pet dogs has been justified through (in chronological order): private entertainment of guests, character building of the adolescent owner, a façade of utility (obedience trials) and concerns of public safety and hygiene. Yet these reasons do not seem sufficient in the contemporary world, where the affective connection with the dog is most certainly the primary one. The dog's pleasure and happiness are the most valid justifications for training nowadays and training – even if it is called behavior modification – addresses the dog as a subject of a new affirmative biopolitics. It is about making the dog's life better. The human's life also improves (as in the slogan: "My rescue dog rescued me"), but this is in the background: the explicit ethical grounding of training is in improving the dog's life. Developments in training are guided by various indicators of the dog's well-being and happiness and the discussion within the dog training community concerns these indicators: how do we make sure that the activity we are engaging in with our dog is beneficial for him? Do the benefits of releasing the urban dog's pent-up energy outweigh the potential for injury associated with agility or flyball? Can the "affective labor" of therapy dogs be justified? How can we develop protocols that ensure that these dogs are not harmed as they help others? These are the kinds of questions that preoccupy contemporary trainers. Of course, there are exceptions: stories of abuse and exploitation, but these do not reverse the general direction of changes. Recognition of the primacy of the affective relationship between human and canine is the guiding principle of these changes. Training becomes a supplement to the human-canine relationship: a supplement that, we like to believe, allows us to improve our dog's life at the same time as it improves ours. It allows us to learn more about our dog; to understand him better. It also becomes a challenge for the human, forcing him or her to reconsider human uniqueness;

to transcend humanity. Is this likely to change? It might. In each and every period under consideration in this book, training performed the dominant take on the human-canine relationship and has performed it as "only natural." It has performed human benevolence and canine sagacity; canine usefulness to man; canine submission to human superior strength and intelligence and, finally, human/canine pleasure and mutual understanding. The fantasy we are emerged in now – and one which is not unique to dog training – is of the potential bliss of interspecies collaboration.

References

Agamben, Giorgio. 1988. *Homo Sacer: Sovereign Power and Bare Life*. Translated by Daniel Heller-Roazen. Stanford, CA: Stanford University Press.

Ailsby, Sue. 2011. "Training Levels. Steps to Success. Level One." *Mind to Mind Publishing. Sue-Eh.Com*. Accessed Oct. 10 2014. http://sue-eh.ca/page24/page26/styled/

Alger, Janet M., and Steven F. Alger. 2013. "Canine Soldiers, Mascots and Stray Dogs in U.S. Wars: Ethical Considerations." In *Animals and War: Studies of Europe and North America*, edited by Ryan Hediger, 45–75. Leiden, Netherlands: Brill.

American Humane Association. 2006. "'Dog Whisperer' Methods More Harmful Than Helpful." *American Humane Association*, September 16, 2006, accessed Oct. 15 2017. http://www.americanhumane.org/about-us/newsroom/news-releases/06-dog-whisperer.html

A Visit to the Circus, Wonderful Performing Dogs, the Rival Musicians and Other Stories. 1899. Philadelphia: Franklin Book Company.

Bailey, Robert, and Marian Bailey. 2012. "The Best Animal Trainers in History: Interview with Bob and Marian Bailey by Sophia Yin, Part I." *Cattledog Publishing: The Art and Science of Animal Behavior* (blog). August 13, accessed April 4, 2017. https://drsophiayin.com/blog/entry/the-best-animal-trainers-in-history-interview-with-bob-and-marian-bailey/

Barad, Karen. 2003. "Posthumanist Performativity: Toward an Understanding of How Matter Comes to Matter." *Signs: Journal of Women in Culture and Society* 28, no. 3: 801–31.

Baratay, Eric. 2012. *Le point de vue animal*. Paris: Le Seuil.

Bederman, Gail. 2008. *Manliness and Civilization: A Cultural History of Gender and Race in the United States, 1880–1917*. Chicago: University of Chicago Press.

Bennett, Pauleen Charmayne, and Vanessa Ilse Rohlf. 2007. "Owner-Companion Dog Interactions: Relationships Between Demographic Variables, Potentially Problematic Behaviours, Training Engagement and Shared Activities." *Applied Animal Behaviour Science* 102, no. 1: 65–84. DOI: 10.1016/j.applanim.2006.03.009.

Berlant, Lauren. 1997. *Queen of America Goes to Washington City: Essays on Sex and Citizenship*. Durham, NC: Duke University Press.

Berns, Gregory. 2013. *How Dogs Love Us: A Neuroscientist and His Adopted Dog Decode the Canine Brain*. Boston & New York: New Harvest.

Biggle, Jacob. 1900. *Biggle Pet Book: A Collection of Information for Old and Young Whose Natural Instincts Teach Them to be Kind to All Living Creatures*. Philadelphia: Wilmer Atkinson.

Birke, Lynda, Mette Bryld, and Nina Lykke. 2004. "Animal Performances: An Exploration of Intersections between Feminist Science Studies and Studies of Human/Animal Relationships." *Feminist Theory* 5, no. 2: 167–183. DOI: 10.1177/1464700104045406.

Birke, Lynda, and Keri Brandt. 2009. "Mutual Corporeality: Gender and Human/Horse Relationships." *Women's Studies International Forum* 32, no. 3 (May/June): 189–197. DOI: 10.1016/j.wsif.2009.05.015.

Bjork, Daniel W. 1993. *B.F. Skinner: A Life.* New York: Basic Books.

Blackwell, Emily J, Caroline Twells, Anne Seawright, and Rachel A. Casey. 2008. "The Relationship Between Training Methods and the Occurrence of Behavior Problems, as Reported by Owners, in a Population of Domestic Dogs." *Journal of Veterinary Behavior: Clinical Applications and Research* 3, no. 5 (September): 207–217. DOI: 10.1016/j.jveb.2007.10.008.

Boddice, Rob. 2012. "The Historical Animal Mind: 'Sagacity' in Nineteenth-Century Britain." In *Encountering Animal Minds: An Anthology of Animal-Human Encounters*, edited by Julie A. Smith and Robert W. Mitchell, 65–78. New York: Columbia University Press.

Bogardus, Adam H. 1874. *Field, Cover and Trap Shooting. Embracing Hints for Skilled Marksmen; Instructions for Young Sportsmen; Haunts and Habits of Game Birds; Flight and Resorts of Water Fowl; Breeding and Breaking of Dogs*. Edited by Charles J. Foster. New York: J. B. Ford & Company.

Bradshaw, John. W. S., Emily J. Blackwell, and Rachel A. Casey. 2009. "Dominance in Domestic Dogs: Useful Construct or Bad Habit?" *Journal of Veterinary Behavior* 4, no. 3 (May/June): 135–144. DOI: 10.1016/j.jveb.2008.08.004.

Braidotti, Rosi. 2002. *Metamorphoses: Towards a Materialist Theory of Becoming*. Malden, MA and Cambridge, UK: Polity.

Breland, Keller, and Marian Breland. 1961. "The Misbehavior of Organisms." *American Psychologist* 16, no. 11: 681–684. DOI: 10.1037/h0040090.

Brockwell, David. 1924. *The Police Dog*. New York: G. Howard Watt.

Bryce, Clarence Archibald. 1909. *The Gentleman's Dog: His Rearing, Training and Treatment*. Richmond, VA: Southern Print Clinic.

Burch, Mary R., Duane Pickel. 1990. "A Toast to Most, a 1910 Pioneer in Animal Training." *Journal of Applied Behavior Analysis* 23, no. 2 (Summer): 263–264.

Burch, Mary R., and Jon S. Bailey. 1999. *How Dogs Learn*. New York: Howell Book House.

Butler, Francis. 1877 [1857]. *Breeding, Training, Management, Diseases of Dogs*. Brooklyn: D.S. Holmes.

Butler, Judith. 1990. *Gender Trouble: Feminism and the Subversion of Identity*. London: Routledge.

Butler, Judith. 1993. *Bodies that Matter: On the Discursive Limits of 'Sex'*. London: Routledge.

Caius, John. 1912. *De Cannibus Britannicis/Of English Dogges*. Translated by Abraham Fleming. *Project Gutenberg*, 2008. http://www.gutenberg.org/files/27050/27050-h/27050-h.htm

Campbell, John. 1994. "'My Constant Companion': Slaves and Their Dogs in the Antebellum South." In *Working Toward Freedom: Slave Society and Domestic Economy in the American South*, edited by Larry E. Hudson, Jr., 53–76. Rochester, NY: University of Rochester Press.

Capshew, James H. 1996. "Engineering Behavior: Project Pigeon, World War II, and the Conditioning of B.F. Skinner." In *B.F. Skinner and Behaviorism in American Culture*, edited by Laurence D. Smith and William R. Woodward, 128–150. Bethlehem: Lehigh University Press.

Carlton, H.W. 1945 [1915]. *Spaniels: Their Breaking for Sport and Field Trials.* Washington, D.C.: Mile G. Delinger.

Cassidy, Rebecca. 2006. "'I Want to Know About the Dogs': Review of *The Companion Species Manifesto: Dogs People and Significant Otherness* by Donna Haraway." *Theory, Culture & Society* 23, no. 7–8: 324–328.

Cesar 911. 2014. Season 1, Episode 1. "Family Feud." Aired March 7, Net Geo Wild.

Cesar 911. Season 1, Episode 2. 2014. "Boss with a Bark." Aired March 14, Nat Geo Wild.

Cesar 911s. Season 1, Episode 5. 2014. "Ripped Apart." Aired April 4, Nat Geo Wild.

Cisney, Vernon W., and Nicolae Morar. 2016. "Why Biopower? Why Now?" *Biopower: Foucault and Beyond*, edited by Vernon W. Cisney and Nicolae Morar, 2–25. Chicago and London: University of Chicago Press.

Chapman, Samuel G. 1990 [1972]. *Police Dogs in North America.* Springfield, IL: Charles C. Thomas.

Chrulew, Matthew, and Dinesh Joseph Wadiwel, eds. 2016. *Foucault and Animals.* Leiden and Boston: Brill.

Cleaver, Elinor R. 1929. "Muzzle or Leash?" *New York Herald Tribune*, Nov. 30, 18. Proquest Historical Newspapers Database.

Clough, Patricia Ticineto. 2003. "Affect and Control: Rethinking the Body 'Beyond Sex and Gender'." *Feminist Theory* 4, no. 3: 359–364.

Cochran, Ford. 2010. "Cesar Millan Shares Rules for Raising Better Dogs." *National Geographic. Voice: Ideas and Insights from Explorers.* October 5, accessed April 5, 2017. https://blog.nationalgeographic.org/2010/10/05/cesar-millan-shares-rules-for-raising-better-dogs/

Cooper, Courtney Ryley. 1928. *Tigers 'N' Lions 'N' Everything.* Boston: Little Brown & Co.

Coppinger, Raymond, and Lorna Coppinger. 2001. *Dogs: A New Understanding of Canine Behavior, Origin and Evolution.* Chicago: University of Chicago Press.

Coppinger, Raymond, and Lorna Coppinger. 2016. *What is a Dog?* Chicago: University of Chicago Press.

Coppinger, Raymond, and Mark Feinstein. 2015. *How Dogs Work.* Chicago: University of Chicago Press.

Coren, Stanley. 1999. "Psychology Applied to Animal Training." In *Psychology: Fields of Application,* edited by Astrid M. Stec and Douglas A. Bernstein, 199–217. Boston, MA: Houghton Mifflin.

Coren, Stanley. 2016a. "Are Women Better Dog Trainers Than Men?" Psychology Today, Jan 21, accessed March 2, 2018. https://www.psychologytoday.com/blog/canine-corner/201601/are-women-better-dog-trainers-men

Coren, Stanley. 2016b. "The Data Says: Don't Hug Your Dog," *Psychology Today,* April 13, accessed April 11, 2017. https://www.psychologytoday.com/blog/canine-corner/201604/the-data-says-dont-hug-the-dog

Cutt, Hayle, Billie Giles-Corti, Matthew Knuiman, Anna Timperio, and Fiona Bull. 2008. "Understanding Dog Owners' Increased Levels of Physical Activity: Results from RESIDE." *American Journal of Public Health* 98, no.1 (January): 66–69. DOI: 10.2105/AJPH.2006.103499.

Darnton, Robert. 2009 [1984]. *The Great Cat Massacre and Other Episodes in French Cultural History*. New York: Basic Books.

Darwin, Charles. 1859. On the Origin of Species. London: John Murray.

Darwin, Charles.1981 [1872]. *The Descent of Man.* Princeton, NJ: Princeton University Press.

Davis, Janet M. 2002. *The Circus Age: Culture & Society Under the American Big Top.* Charlotte: University of North Carolina Press.

Deleuze, Gilles, and Felix Guattari. 1987. *A Thousand Plateaus: Capitalism and Schizophrenia.* Minneapolis: University of Minnesota Press.

Deleuze, Gilles. 1992. "Postscript on the Societies of Control." *October* 59 (Winter): 3–7.

Derr, Mark. 2004. *A Dog's History of America.* New York: North Point Press.

Derr, Mark. "A Pack of Lies." 2006. *The New York Times*, August, 31 2006, accessed April 25, 2017. http://www.nytimes.com/2006/08/31/opinion/31derr.html

Desmond, Jane.1999. *Staging Tourism: Bodies on Display from Waikiki to Sea World.* Chicago: University of Chicago Press.

De Waal, Frans. 1996. *Good Natured: The Origins of Right and Wrong in Humans and Other Animals.* Cambridge and London: Harvard University Press.

De Waal, Frans. 2010. *The Age of Empathy: Nature's Lessons for a Kinder Society.* New York: Broadway Books.

Dickey, Bronwen. 2016. *Pit Bull: The Battle over an American Icon.* New York: AA Knopf.

Dobson, Michael. 2000. "A Dog at All Things: The Transformation of the Onstage Canine, 1550–1850." *Performance Research* 5, no. 2: 116–124. DOI: 10.1080/13528165.2000.10871736.

Dodge, Geraldine, and Josephine Z. Rine. 1956. *The German Shepherd Dog in America.* New York: Orange Judd Publishing Company.

Donaldson, Jean. 1996. *The Culture Clash: A New Way of Understanding the Relationship Between Humans and Domestic Dogs.* Berkeley, CA: James and Kenneth Publishers.

Donaldson, Jean. 2016. "Interview with Jean Donaldson on *The Culture Clash* by Zazie Todd." *Companion Animal Psychology*. November 16, accessed April 18, 2017. http://www.companionanimalpsychology.com/2016/11/interview-with-jean-donaldson-on.html

Donaldson, Sue, and Will Kymlicka. 2011. *Zoopolis: A Political Theory of Animal Rights*. Oxford: Oxford University Press.

Downey, Fairfax Davis. 1955. *Dogs for Defense: American Dogs in the Second World War, 1941–45*. N.p.: Daniel P. McDonald.

Farrell, Jocelyn M., Hope, Ashley E. Hope, Rodney Hulstein, and Sandi J. Spaulding. 2015. "Dog-Sport Competitors: What Motivates People to Participate with Their Dogs in Sporting Events?" *Anthrozoös 28*, no. 1: 61–71. DOI: 10.2752/089279315X14129350722010

Fenzi, Denise. 2016. *Train the Dog in Front of You*. N.p.: Fenzi Dog Sports Academy.

Fisher, Gail Tamases. 2009. *The Thinking Dog: Crossover to Clicker Training*. Wenatchee, WA: Dogwise.

Fornet-Betancourt, Raul, Helmut Becker, and Alfredo Gomez-Muller. 1987. "The Ethic of Care for the Self as a Practice of Freedom: An Interview with Michel Foucault on January 20, 1984." Translated by J.D. Gauthier. *Philosophy & Social Criticism* 12, no.2–3: 112–131.

Forester, Frank. 1856. *The Complete Manual for Young Sportsmen: with Directions for Handling the Gun, the Rifle and the Rod; the Art of Shooting on the Wing; the Breaking, Management and Hunting of the Dog; the Varieties and Habits of Game; River, Lake and Sea Fishing, Etc*. New York: Geo E. Woodward.

Foucault, Michel. 1978. *History of Sexuality. Volume 1: An Introduction*. Translated by Robert Hurley. New York: Pantheon.

Foucault, Michel. 1980. *Power/Knowledge: Selected Interviews and Other Writings, 1972–1977*. Edited by Colin Gordon. Brighton: Harvester.

Foucault, Michel. 1982. "The Subject and Power." *Critical Inquiry* 8, no. 4 (Summer): 777–795.

Foucault, Michel. 1984a. "Nietzsche, Genealogy, History." In *The Foucault Reader*, edited by Paul Rabinow, 76–100. New York: Pantheon.

Foucault, Michel. 1984b. "On the Genealogy of Ethics: A Work in Progress." In *The Foucault Reader*, edited by Paul Rabinow, 340–372. New York: Pantheon.

Foucault, Michel. 1984c. "Truth and Power." In *The Foucault Reader,* edited by Paul Rabinow, 51–75. New York: Pantheon.

Foucault, Michel. 1984d. "What is Enlightenment?" In *The Foucault Reader,* edited by Paul Rabinow, 32–50. New York: Pantheon.

Foucault, Michel. 1985. *The Use of Pleasure: The History of Sexuality, Volume 2*. Translated by R. Hurley. New York: Pantheon.

Foucault, Michel. 1988. *Technologies of the Self: A Seminar with Michel Foucault*. Edited by Luther H. Martin, Huck Gutman, and Patrick H. Hutton. Amherst: University of Massachusetts Press.

Foucault, Michel. 1994. *The Order of Things: An Archeology of the Human Sciences*. New York: Vintage.

Foucault, Michel. 1995. *Discipline and Punish: The Birth of the Prison*. Translated by Alan Sheridan. New York: Vintage.

Foucault, Michel. 2001. *Fearless Speech*. Edited by Joseph Pearson. Los Angeles, CA: Semiotexte.

Foucault, Michel. 2003. *Society Must Be Defended: Lectures at the College de France, 1975–76*. Edited by Mauro Bertani and Alessandro Fontana. Translated by David Macey. New York: Palgrave Macmillan/Picador.

Foucault, Michel. 2007. *Security, Territory, Population: Lectures at the College de France, 1977/78*. Edited by Michel Senellart. Translated by Graham Burchell. New York: Palgrave Macmillan/Picador.

Foucault, Michel. 2008. *The Birth of Biopolitics: Lectures at the College de France, 1978–1979*. Edited by Michel Sennelart. Translated by Graham Burchell. New York: Palgrave MacMillan/Picador.

Foucault, Michel. 2012. *The Courage of Truth: Lectures at the College de France 1983–1984*. Edited by Frederic Gros. Translated by Graham Burchell. New York: Palgrave MacMillan/Picador.

Foucault, Michel. 2013. *History of Madness*. Edited by Jean Khalfa. Translated by Jonathan Murphy and Jean Khalfa. New York: Routledge.

Franklin, Adrian. 1999. *Animals and Modern Cultures: A Sociology of Human-Animals Relations in Modernity*. London: Sage Publications.

Fudge, Erica. 2002. "A Left-Handed Blow: Writing the History of Animals." In *Representing Animals*, edited Nigel Rothfels, 3–18. Bloomington: Indiana University Press.

Fudge, Erica. 2008. *Pets*. New York and London: Routledge.

Fugazza, Claudia. 2014. *Do As I Do: Using Social Learning to Train Dogs*. Wenatchee, WA: Dogwise.

Garrett, Susan. 2002. *Ruff Love: A Relationship Building Program for You and Your Dog*. Chicopee, MA: Clean Run Productions.

Garrett, Susan. 2005. *Shaping Success: The Education of an Unlikely Champion*. Chicopee, MA: CleanRun Productions.

Garrett, Susan. 2007. *Crate Games for Motivation and Control*. Alberton, ON: Say Yes! Dog Training. DVD.

Geertz, Clifford. 2005 [1973]. "Deep Play: Notes on the Balinese Cockfight." *Daedalus* 134, no. 4 (Fall): 56–86.

Gersbach, Robert, and Theo F. Jager. 1910. *The Police Dog in Word and Picture: A Complete History of Police Dogs [...] the Trainer's Handbook, the Breeder's Guide, the Officer's Vade Mecum*. Canandaigue, NY: The Deming Press.

Gillespie, Dair L., Ann Leffler and Elinor Lerner. 2002. "If It Weren't For My Hobby, I'd Have a Life: Dog Sports, Serious Leisure, and Boundary Negotiations." *Leisure Studies* 21: 285–304

Gilmette, Lauren. 2014. "In What We Tend to Feel is Without History: Foucault, Affect and the Ethics of Curiosity." *The Journal of Speculative Philosophy* 28, no. 3: 284–294. DOI: 10.5325/jspecphil.28.3.0284.

Giltner, Scott E. 2008. *Hunting and Fishing in the New South: Black Labor and White Leisure After the Civil War*. Baltimore: The Johns Hopkins University Press.

Glenney Boggs, Coleen. 2013. *Animalia Americana: Animal Representations and Biopolitical Subjectivity*. New York: Columbia University Press.

Gorant, Jim. 2010. *The Lost Dogs: Michael Vick's Dogs and Their Tale of Rescue and Redemption*. New York: Penguin.

Gorant, Jim. 2012. *Wallace: The Underdog Who Conquered a Sport, Saved a Marriage, and Championed Pit Bulls–One Flying Disc at a Time*. New York: Gotham.

Gorant, Jim. 2017. *The Found Dogs: The Fates and Fortunes of Michael Vick's Pitbulls, 10 Years After Their Heroic Rescue*. N.p.: BookBaby.

Gordon, Colin. 1991. "Governmental Rationality: An Introduction." In *The Foucault Effect: Studies in Governmentality*, edited by Graham Burchell, Colin Gordon, and Peter Miller, 1–53. Chicago: The University of Chicago Press.

Grier, Katherine. 2006. *Pets in America: A History*. Charlotte: University of North Carolina Press.

Gros, Frederic. 2016. "Is There a Biopolitical Subject? Foucault and the Birth of Biopolitics," translated by Samantha Bankston. In *Biopower: Foucault and Beyond*, edited by Veron W. Cisney and Nicolae Morar, 259–273. Chicago and London: The University of Chicago Press.

Hadra, Andrew C., dir. 2012. *Away to Me: The Sheepdog Movie*. Main Street Partners, DVD.

Hagenbeck, Carl. 1909. Beasts and Men. Translated by Hugh S. R. Elliot and A.G. Thacker. London, New York, Calcutta: Longmans, Green & Co.

Hammond, Stephen Tillinghast. 1882. *Practical Dog Training, or Training Vs. Breaking*. New York: Forest and Stream.

Haney, Jesse. 1869. *Haney's Art of Training Animals: A Practical Guide for Amateur and Professional Trainers, Giving Full Instructions for Breaking, Taming and Teaching All Kinds of Animals*. New York: Jesse Haney & Co, Publishers.

Haney, Jesse. 1894. *How to Train Animals*. The People's Handbook Series. New York: F. M. Lupton.

Haraway, Donna. 1984. "Teddy Bear Patriarchy: Taxidermy in the Garden of Eden, New York City, 1908–1936." *Social Text* 11 (Winter): 20–64. DOI: 10.2307/466593

Haraway, Donna. 1989. *Primate Visions*. New York: Routledge.

Haraway, Donna. 2003. *Companion Species Manifesto*. Chicago: Prickly Paradigm Press.

Hara ay, Donna. 2008. *When Species Meet*. Minneapolis: University of Minnesota Press.

Hardt, Michael. 2010. "Militant Life." *New Left Review* 64 (July/August): 151–160.

Hardt, Michael, and Antonio Negri. 2001. *Empire*. Cambridge, MA: Harvard University Press.

Hare, Brian, and Vanessa Woods. 2013. *The Genius of Dogs: Discovering the Unique Intelligence of Man's Best Friend.* New York: Penguin.

Havas, Randall. 2012. "Individuality as Reliability: A Dog Trainer's Guide to Nietzsche." *Journal of Nietzsche Studies* 43, no. 1 (Spring): 18–31. DOI: 10.5325/jnietstud.43.1.0018.

Hawker, Peter. 1853 [1846]. *Instructions to Young Sportsmen, in All That Relates to Guns and Shooting.* Philadelphia: Blanchard and Lea.

Hayles, N. Katherine. 2002. "Flesh and Metal: Reconfiguring the Mindbody in Virtual Environments." *Configurations* 10, no. 2: 297–320. DOI: 10.1353/con.2003.0015.

Hearne, Vicki. 1986. *Adam's Task: Calling Animals by Name.* New York: Alfred A. Knopf.

Hearne, Vicki. 1991. "What's Wrong with Animal Rights? Of Hounds, Horses and Jeffersonian Happiness." *Harper's Magazine* 283 (September): 59–64.

Hearne, Vicki. 1995. *Animal Happiness.* New York: Harper Perennial.

Hedrick, Donald. 2012. "Too Much Fun: Shakespeare's Entertainment Unconscious." *Journal of the Midwest Modern Language Association* 45, no. 2 (Fall): 17–39. DOI: 10.1353/mml.2012.0022

Hocutt, Max. 2013. "The Fruits and Fallacies of Fred Skinner on Freedom." *The Independent Review* 18, no. 2: 263–278.

Homans, John. 2012. *What's a Dog For? The Surprising History, Science, Philosophy, and Politics of Man's Best Friend.* New York: Penguin Press.

House and Pet Dogs: Their Selection, Care and Training. 1890. New York: Forest and Stream Publishing.

Howell, Philip. 2015. *At Home and Astray: The Domestic Dog in Victorian Britain.* Richmond, VA: University of Virginia Press.

Hughes, Holly. 2015. "It's Hard to Spot the Queerness in this Image." In *Performing Animality: Animals in Performance Practices*, edited by Lourdes Orozco and Jennifer Parker-Starbuck, 204–217. New York and London: Palgrave Macmillan.

Hultsman, Wendy Z. 2012. "Couple Involvement in Serious Leisure: Examining Participation in Dog Agility." *Leisure Studies* 31, no. 2 (2012): 231–253. DOI: 10.1080/02614367.2011.619010.

Irvine, Leslie and Jenny R. Vermilya. 2010 "Gender Work in a Feminized Profession: The Case of Veterinary Medicine." *Gender & Society* 24, no. 1: 56–82. DOI: 10.1177/0891243209355978

Jager, Theo F. 1917. *Scout, Red Cross and Army Dogs: A Historical Sketch of Dogs in the Great War and a Training Guide for the Rank and File of the United States Army*. Rochester, NY: Arrow Printing Company.

Jackson-Schebetta, Lisa. 2009. "Mythologies and Commodifications of Dominion in *The Dog Whisperer* with Cesar Millan." *Journal for Critical Animal Studies* 7, no. 1: 107–131.

Jacobs, Debbie. 2008. *A Guide to Living with and Training a Fearful Dog*. Wenatchee, WA: Dogwise.

Jennings, John J. 1882. *Theatrical and Circus Life; or Secrets of the Stage Room, Green-Room and Sawdust Arena*. St. Louis: Sun Publishing.

K9 Nosework. 2014. "What is K9 Nosework?" *K9nosework.com*, accessed September 9, 2016. http://www.k9nosework.com/about-us/what-k9-nose-work

Kean, Hilda. 1998. *Animal Rights: Political and Social Change in Britain since 1800*. London: Reaktion.

Kete, Kathleen. 1994. *The Beast in the Boudoir: Petkeeping in Nineteenth-Century Paris*. Berkley, Los Angeles, London: University of California Press.

Kim, Claire Jean. 2015. *Dangerous Crossings: Race, Species and Nature in a Multicultural Age*. New York: Cambridge University Press.

Kimmel, Michael. 2001. "Masculinity as Homophobia: Fear, Shame, and Silence in the Construction of Gender Identity." In *The Masculinities Reader*, edited by Stephen Whitehead, 266–287. New York: Polity Press.

Koehler, William. 1976 [1962]. *The Koehler Method of Dog Training: Certified Techniques by Movieland's Most Experienced Dog Trainer*. New York: Howell Book House.

Kollet, Fred. 1926 [1924]. *Training the Police Dog*. Chicago: Judy Publishing Company.

Kramer, Charles L. 2003. "Introduction." *Rally-O: An Introduction to Rally Style Obedience*, accessed May 15 2017. http://www-personal.ksu.edu/~kramerc/intro-to-2nd-ed.htm

La Caze, Marguerite, and Henry Martyn Lloyd. 2011. "Editor's Introduction: Philosophy and the 'Affective Turn.'" *Parrhesia* no. 13: 1–13.

Lane, Marion S., and Stephen L. Zawistowski. 2008. *Heritage of Care: The American Society for the Prevention of Cruelty to Animals*. Westport, CT and London: Praeger Books.

Lewis, Elisha J. 1855. *American Sportsman: Containing Hints to Sportsmen, Notes on Shooting and the Habits of the Game Birds and Wild Fowl of America*. Philadelphia: Lippincott, Grambo and Co.

Lewis, Tania. 2008. *Smart Living: Lifestyle Media and Popular Expertise*. New York: Peter Lang.

Life. 1940. "How to Teach a Dog Manners is Shown by Pretty Girl Trainer," June 17: 68–73.

Lindsay, Steven R. 2001. "History of Applied Dog Behavior and Training," In *Handbook of Applied Dog Behavior and Training: Etiology and Assessment of Behavior Problems, Volume 2*, 3–24. Ames, IA: Iowa State University Press.

Loeb, Paul S. 2017. "Nietzsche's Place in the Aristotelian History of Philosophy." In *Nietzsche and the Philosophers*, edited by Mark T. Conrad. New York: Routledge. Kindle Ebook.

Logan, Olive. 1871. *The Mimic World and Public Exhibitions: Their History, Their Morals, and Effects*. Philadelphia and Cincinnati: New World Publishing Company.

Look. 1952. "Harvard Trained Dog." May 20: 17–20.

London, Jack. 1906. White Fang. New York: Macmillan.

Lord, Kathryn, Richard Schneider, and Raymond Coppinger. 2016. "The Evolution of Working Dogs" In *The Domestic Dog: Its Evolution, Behavior and Interactions with People. Second Edition*, edited by James Serpell, 42–66. Cambridge: Cambridge University Press.

Lorenz, Konrad. 1952. *King Solomon's Ring: New Light on Animal Ways*. Translated by Marjorie Kerr Wilson. London: Methuen.

Lorenz, Konrad. 1953. *Man Meets Dog*. Translated by Marjorie Kerr Wilson. Baltimore: Penguin Books.

Lott, Eric. 1993. *Love and Theft: Blackface Minstrelsy and the American Working Class*. New York and Oxford: Oxford University Press.

Lou, Jo Anna. 2017. "Try A New Dog Sport This Summer." *The Bark,* Apr. 27, accessed. May 11, 2017. http://thebark.com/content/try-new-dog-sport-summer

Luescher, Andrew Urs, and Robert Tyson Medlock. 2009. "The Effects of Training and Environmental Alterations on Adoption Success of Shelter Dogs." *Applied Animal Behaviour Science* 117, no. 1 (February): 63–68. DOI: 10.1016/j.applanim.2008.11.001.

Lund, Giuliana. 2014. "Taking Teamwork Seriously: The Sport of Dog Agility as an Ethical Model of Cross-Species Companionship." In *Sport, Animals, and Society*, edited by James Gillett and Michelle Gilbert, 101–123. New York and London: Routledge/

Mackintosh, Alex. 2016. "Foucault's Menagerie: Cock-Fighting, Bear Baiting, and the Genealogy of Human-Animal Power." In *Foucault and Animals*, edited by Matthew Chrulew and Dinesh Joseph Wadiwel, 161–189. Boston and Leiden: Brill.

Marks, Stuart A. 1991. *Southern Hunting in Black and White: Nature, History and Ritual in a Carolina Community*. Princeton, NJ: Princeton University Press.

Marr, John N. 2002. "Marian Breland Bailey: The Mouse Who Reinforced." *The Arkansas Historical Quarterly* 61, no. 1 (Spring): 59–79. DOI: 10.2307/40031038.

Marsh, Mary More. 1893. "A Trick Dog: A Dog that Learned Geometry by Training." *Los Angeles Times*, Sep. 24: 14. ProQuest Historical Newspapers.

Marvin, Garry. 2000. "Natural Instincts and Cultural Passions: Transformations and Performances in Foxhunting." *Performance Research* 5, no. 2: 108–115. DOI: 10.1080/13528165.2000.10871735.

Marvin, Garry. 2001. "Cultured Killers: Creating and Representing Foxhounds." Society & Animals 9, no. 3 : 273-292. DOI: 10.1163/156853001753644417.

Marvin, Garry. 2003. "A Passionate Pursuit: Foxhunting as Performance." *The Sociological Review* 51, no. 2 (October): 46–60. DOI: 10.1111/j.1467-954X.2004.00450.x.

Marx, Leo. 1964. *The Machine in the Garden: Technology and the Pastoral Ideal in America*. London, Oxford and New York: Oxford University Press.

Marx, Leo. 1986. "Pastoralism in America." In *Ideology and Classic American Literature*, edited by Sacvan Bercovitch and Myra Jehlen, 36–69. Cambridge, MA: Cambridge University Press.

Mason, Jennifer. 2005. *Civilized Creatures: Urban Animals, Sentimental Culture, and American Literature, 1850–1900*. Baltimore: The Johns Hopkins University Press.

Massumi, Brian. 1987. "Notes on the Translation and Acknowledgements." In *A Thousand Plateaus: Capitalism and Schizophrenia*, by Gilles Deleuze and Felix Guattari, xvi–xix. Minneapolis: University of Minnesota Press.

McCaig, Donald. 1984. *Nop's Trials*. New York: Crown Publishers.

McCaig, Donald. 1991. *Eminent Dogs and Dangerous Men*. New York: Harper Collins.

McCaig, Donald. 2007a. *A Useful Dog*. Charlottsville, VA: University of Virginia Press.

McCaig, Donald. 2007b. *The Dog Wars: How the Border Collie Battled the American Kennel Club*. Kearney, NE: Outrun Press.

McCaig, Donald. *Mr. and Mrs. Dog: Our Travels, Trials, Adventures and Epiphanies*. Charlottsville, VA: University of Virginia Press, 2013.

McCarthy, Anna. 2007. "Reality Television: A Neoliberal Theater of Suffering." *Social Text* 25, no. 4 (Winter): 17–41. DOI: 10.1215/01642472-2007-010.

McConnell, Patricia. 2002. *The Other End of the Leash: Why We Do What We Do Around Dogs*. New York: Ballantine Books.

McHugh, Susan.2004. *Dog*. London: Reaktion Books.

McWhorter, Diane. 2001. *Carry Me Home: Birmingham, Alabama: The Climactic Battle of the Civil Rights Revolution*. New York: Simon and Schuster.

Millan, Cesar, and Melissa Jo Peltier. 2007. *Cesar's Way: The Natural, Everyday Guide to Understanding and Correcting Common Dog Problems*. New York: Harmony Books.

Miller, Pat. 2001. *The Power of Positive Dog Training*. New York: Howell Publishing.

Mitchell, Robert W., and Mark Hamm. 1997. "The Interpretation of Animal Psychology: Anthropomorphism or Behavior Reading?" *Behaviour* 134, no. 3: 173–204.

Mizelle, Brett. 2005. "Contested Exhibitions: The Debate over Proper Animal Sights in Post-Revolutionary America." *Worldviews: Global Religions, Culture, and Ecology* 9, no. 2: 219–235. DOI: 10.1163/1568535054615367.

Monks of New Skete. 1978. How to Be Your Dog's Best Friend. New York: Little, Brown & Company.

Most, Konrad. 1955. *Training Dogs: A Manual*. Translated by James Cleugh. New York: Coward-McCann, Inc., 1955.

Murray, Susan, and Laurie Ouellette, eds. 2004. *Reality TV: Remaking Television Culture*. New York: NYU Press.

Nance, Susan. 2013. *Entertaining Elephants: Animal Agency and the Business of the American Circus.* Baltimore: The Johns Hopkins University Press.

Nealon, Jeffrey. 2008. *Foucault Beyond Foucault: Power and Its Intensifications since 1984*. Stanford, CA: University of California Press.

Nealon, Jeffrey. 2016. *Plant Theory: Biopower & Vegetable Life*. Stanford, CA: Stanford University Press.

Neuringer, Allen, and Walter Englert. 2017. "Epicurus and BF Skinner: In Search of the Good Life." *Journal of the Experimental Analysis of Behavior* 107, no. 1: 21–33. DOI: 10.1002/jeab.230.

New York Times. 1871. "Training Dogs: A Morning Chat with a Professional Dog Teacher." January 22, 8. ProQuest Historical Newspapers Database.

New York Times. 1878. "The End of the Dog Show. Closing Scenes in the Garden. A Fox Hunt and a Field Trial…" May 19, 2. ProQuest Historical Newspapers Database.

New York Times. 1913. "Police Dogs Give Exhibition in Park." August 24, 5. Proquest Historical Newspapers Database.

Nietzsche, Friedrich. 1921. *The Genealogy of Morals.* Translated by Horace B. Samuel. New York: Boni and Liveright.

Nimmo, Richie. 2010. *Milk, Modernity and the Making of the Human: Purifying the Social*. New York: Routledge.

Orlean, Susan.2011. *Rin Tin Tin: The Life and the Legend.* New York: Simon and Schuster.

Orłowska, Agnieszka. 2016. "Towards Mutual Understanding, Respect, and Trust: On Past and Present Dog Training in Poland." In *Free-Market Dogs: The Human-Canine Bond in Post-Communist Poland,* edited by Michał P. Pręgowski and Justyna Włodarczyk, 35–60. West Lafayette, IN: Purdue University Press, 2016.

Oulette, Laurie. 2009. "Take Responsibility for Yourself: *Judge Judy* and the Neoliberal Citizen." In *Reality TV: Remaking Television Culture,* edited by Susan Murray and Laurie Oulette, 223–241. New York and London: New York University Press.

Palmer, Clare. 2001. "'Taming the Wild Profusion of Existing Things?' A Study of Foucault, Power and Human/Animal Relationships." *Environmental Ethics* 23, no. 4 (Winter): 339–359. DOI: 10.5840/enviroethics20012342.

Palmer, Gareth, ed. 2008. *Exposing Lifestyle Television: The Big Reveal.* Hampshire and Burlington, VT: Ashgate Publishing.

Pandian, Anand. 2008. "Pastoral Power in the Postcolony: On the Biopolitics of the Criminal Animal in South India." *Cultural Anthropology* 23, no. 1 (February): 85–117. DOI: 10.1111/j.1548-1360.2008.00004.x.

Patterson, Arthur A. 1944. "Training Class for City's Dogs Proves Popular." *New York Herald Tribune*, July 23, B2. Proquest Historical Newspapers Database.

Patton, Paul. 2003. "Language, Power, and the Training of Horses." In *Zoontologies: The Question of the Animal,* edited by Cary Wolfe, 83–99. Minneapolis: University of Minnesota Press.

Pearl, Michael, and Debi Pearl. 1994. *To Train Up a Child.* Pleasantville, TN: No Greater Joy Ministries.

Pearson, Chris. 2016. "Between Instinct and Intelligence: Harnessing Police Dog Agency in Early Twentieth-Century Paris." *Comparative Studies in Society and History* 58, no. 2 (April): 463–490. DOI: 10.1017/S0010417516000141.

Peterson, Michael. 2007. "The Animal Apparatus: From a Theory of Animal Acting to an Ethics of Animal Acts." *TDR/The Drama Review* 51, no. 1 (Spring): 33–48. DOI: 10.1162/dram.2007.51.1.33.

Pochmara, Anna. 2009. "Are You a Real Man?- The Construction of Hegemonic Masculinity in American Culture." *Polish Journal for American Studies* 2: 127–140.

Pochmara, Anna. 2011. *The Making of the New Negro: Black Authorship, Masculinity, and Sexuality in the Harlem Renaissance.* Amsterdam: Amsterdam University Press.

Pręgowski, Michał P. 2014. "Między mitem a wiedzą empiryczną. Definiowanie relacji międzygatunkowej w szkoleniu psów." *Miscellanea Anthropologica et Sociologica* 15, no. 1: 55–70.

Pręgowski, Michał P. 2015. "Your Dog is Your Teacher: Contemporary Dog Training Beyond Radical Behaviorism." *Society&Animals* 23, no. 6: 525–543. DOI: 10.1163/15685306-12341383.

Pręgowski, Michał P, and Justyna Włodarczyk. 2016. "Researching the Human-Canine Relationship in Democratic Poland." In *Free Market Dogs: The Human-Canine Bond in Post-Communist Poland*, edited by Michał Pręgowski and Justyna Włodarczyk, 1–34. West Lafayette, Indiana: Purdue University Press, 2016.

Probyn, Elspeth. 1996. "Becoming-Horse: Transports in Desire." In *Outside Belongings,* 37–62. Routledge: New York.

Protopopova, Alexandra, and Clive David Lawrence Wynne. 2014. "Adopter-dog Interactions at the Shelter: Behavioral and Contextual Predictors of Adoption." *Applied Animal Behaviour Science* 157 (August): 109–116. DOI: 10.1016/j.applanim.2014.04.007.

Prozorov, Sergei. 2015. "Foucault's Affirmative Biopolitics: Cynic Parrhesia and the Biopower of the Powerless." *Political Theory* 43: 1–23. DOI: 10.1177/0090591715609963.

Pryor, Ashley E. 2012. "Heidegger and the Dog Whisperer: Imagining Interspecies Kindness." In *Animals and the Human Imagination: A Companion to Animal Studies,* edited by Aaron Gross and Anne Vallely, 289–306. New York: Columbia University Press.

Pryor, Karen. 1984. *Don't Shoot the Dog: How to Improve Yourself and Others Through Behavioral Training.* New York: Simon and Schuster.

Pryor, Karen. 1997. "History of Clicker Training I: Karen Pryor's Acceptance Speech: Annual Award for Excellence in the Media, Association for Behavior Analysis, May 23, 1997." *Clickertraining.com*, accessed April 14 2017. https://clickertraining.com/node/153

Pryor, Karen. 2009. *Reaching the Animal Mind: Clicker Training and What It Teaches Us About All Animals.* New York, London, Toronto, Sydney: Scribner.

Rabinow, Paul and Nikolas Rose. 2006 "Biopower Today." *Biosocieties* 1, no. 2: 195–217. DOI: 10.1017/S1745855206040014.

Randall, Henry Stephens. 1863. The Practical Shepherd: A Complete Treatise on the Breeding, Management and Diseases of Sheep. Rochester. NY and Philadelphia: D.D.T. Moore, Union Buildings.

Rasmussen, Claire Elaine. 2011. *The Autonomous Animal: Self-governance and the Modern Subject*. Minneapolis: University of Minnesota Press.

Rasmussen, Claire Elaine. 2016. "Domesticating Bodies: Race, Species, Sex and Citizenship." In *Political Theory and the Animal/Human Relationship*, edited by Judith Grant and Vincent G. Jungkunz, 75–101. Albany, NY: SUNY Press.

Reisner, Ilana. 2016. "The Learning Dog: A Discussion of Training Methods." In *The Domestic Dog: Its Evolution, Behavior and Interactions with People. Second Edition*, edited by James Serpell, 210–226. Cambridge: Cambridge University Press.

Ritvo, Harriet.1987. *The Animal Estate: The English and Other Creatures in the Victorian Age*. Cambridge, MA and London: Harvard University Press.

Ritvo, Harriet. 1997. *The Platypus and the Mermaid and Other Figments of the Classifying Imagination*. Cambridge, MA and London: Harvard University Press.

Romanes, George. 1882. *Animal Intelligence.* London: Kegan Paul, Trench. Hathi Trust Digital Library. https://hdl.handle.net/2027/mdp.39015008599352

Ronchette, Vicki. 2007. *Positive Training for Show Dogs: Building a Relationship for Success.* Wenatchee, WA: Dogwise.

Rose, Nikolas. 1999a. "Control." In *Powers of Freedom: Reframing Political Thought*, 233–73. Cambridge: Cambridge University Press.

Rose, Nikolas. 1999b. *Governing the Soul: The Shaping of the Private Self.* London and New York: Free Association Books.

Rose, Nikolas. 2001. "Biopolitics in the Twenty First Century-Notes for a Research Agenda." *Distinktion: Scandinavian Journal of Social Theory* 2, no. 3: 25–44. DOI: 10.1080/1600910X.2001.9672798.

Rose, Nikolas. 2009. *The Politics of Life Itself: Biomedicine, Power, and Subjectivity in the Twenty-First Century*. Princeton, NJ and Oxford: Princeton University Press.

Rothfels, Nigel. 2002. *Savages and Beasts: The Birth of the Modern Zoo*. Baltimore: JHU Press.

Rutkowska, Małgorzata. 2016. "'A Dog's Life': Pet-Keeping in Canadian and American Animal Autobiographies at the Turn of the Twentieth Century." *Polish Journal for American Studies* 10: 37–48.

Salman, M. D., John G. New, Jr., Janet M. Scarlett, Phillip H. Kass, Rebecca Ruch-Gallie, and Suzanne Hetts. 1998. "Human and Animal Factors Related to Relinquishment of Dogs and Cats in Twelve Selected Animal Shelters in the United States." *Journal of Applied Animal Welfare Science* 1, no. 3: 207–226. DOI: 10.1207/s15327604jaws0103_2.

Saunders, Blanche. 1954. *Dog Obedience: The Complete Guide for Trainers*. New York: Howell.

Saunders, Blanche. 1965 [1946]. *Training You to Train Your Dog*. Garden City, NY: Doubleday.

Saunders, Blanche. 1974. *The Story of Dog Obedience*. New York: Howell Book House.

Schaffer, Michael. 2009. *One Nation Under Dog: Adventures in the New World of Prozac-Popping Puppies, Dog-Park Politics, and Organic Pet Food*. New York: Henry Holt and Company.

Schofield, Grant, Kerry Mummery, and Rebekah Steele. 2005. "Dog Ownership and Human Health-Related Physical Activity: an Epidemiological Study." *Health Promotion Journal of Australia* 16, no. 1: 15–19. DOI: 10.1071/HE05015.

Schultz, Rhona. 1976. "Obedience Training: A New Leash on Life." *Chicago Tribune*, November 3, F1. ProQuest Historical Newspapers Database.

Sdao, Kathy. 2012. *Plenty in Life is Free: Reflections on Dogs, Training and Finding Grace*. Wenatchee, WA: Dogwise.

Serpell, James. 1999. "Guest Editor's Introduction: Animals in Children's Lives." *Society and Animals* 7, no. 2: 87–94.

Serpell, James. "People in Disguise: Anthropomorphism and the Human-Pet Relationship." In *Thinking with Animals: New Perspectives on Anthropomorphism*, edited by Daston Lorraine and Mitman Gregg, 121–36. New York: Columbia University Press.

Serpell, James, and Yuying A. Hsu. 2005. "Effects of Breed, Sex, and Neuter Status on Trainability in Dogs." *Anthrozoös* 18, no. 3: 196–207. DOI: 10.2752/089279305785594135.

Siettou, Christina, Ian M. Fraser, and Rob W. Fraser. 2014. "Investigating Some of the Factors That Influence 'Consumer' Choice When Adopting a Shelter Dog in the United Kingdom." *Journal of Applied Animal Welfare Science* 17, no. 2: 136–147. DOI: 10.1080/10888705.2014.883924.

Shukin, Nicole. 2009. *Animal Capital: Rendering Life in Biopolitical Times*. Minneapolis: University of Minnesota Press.

Shukin, Nicole. 2013. "Security Bonds: On Feeling Power and the Fiction of an Animal Governmentality." *ESC: English Studies in Canada* 39 no. 1 (March): 177–198. DOI: 10.1353/esc.2013.0015.

Skabelund, Aaron. 2008. "Breeding Racism: The Imperial Battlefields of the 'German' Shepherd Dog." *Society and Animals*. 16, no. 4: 354–371. DOI: 10.1163/156853008x357676.

Skinner, B.F. 1938. *The Behavior of Organisms*. New York: Appleton Century-Crofts.

Skinner, B.F. 1948. *Walden Two*. New York: Macmillan Publishing.

Skinner, B.F. 1953. Science and Human Behavior. New York: Simon & Schuster.

Skinner, B.F. 1964. *Interview with Dr. B.F. Skinner by Richard I. Evans, Part 1*. University Park, PA: Penn State Media, 1964. Online Video Resource, accessed March 3 2017. http://search.alexanderstreet.com/view/work/2182647

Skinner, B.F. 1971. *Beyond Freedom and Dignity*. New York: Alfred Knopf.

Skinner, B.F. 1974. *About Behaviorism*. New York: Vintage Books.

Skinner, B.F. 1983. *Enjoy Old Age: A Program of Self-Management*. New York: W. W. Norton & Co.

Skinner, John Stewart. 1845. *The Dog and the Sportsman*. Philadelphia: Lea and Blanchard.

Smith, Morton. 1947. "The Dog and Pony Show." *Circusnews.com*, accessed June 7 2015. Originally published in *Hobbies*, December 1947: 22–25.

Smith, Julie A. 2012. "The Meaning of Energy in Millan's Discourse on Dogs." *Experiencing Animal Minds: An Anthology of Human-Animal Encounters*, edited by Julie A. Smith and Robert W. Mitchell, 142–153. New York: Columbia University Press.

Spector, Morgan. 1999. *Clicker Training for Obedience: Shaping Top Performance Positively*. Waltham, MA: Karen Pryor Clickertraining.

Spiegel, Marjorie. 1996 [1988]. *The Dreaded Comparison: Human and Animal Slavery*. New York: Mirror Books.

Stanescu, James. 2013. "Beyond Biopolitics: Animal Studies, Factory Farms, and the Advent of Deading Life." *PhaenEx* 8, no. 2: 135–160. DOI: 10.22329/p.v8i2.4090.

Stevenson, Edward I. 1882. "Some Hints on Dog Teaching." *Harper's Young People* 3, no. 158: 823. Hathi Trust Digital Library.

Stremming, Sarah. 2017. "I Wish I'd Known." *The Cognitive Canine* (blog), February 15, accessed April 15 2017. https://thecognitivecanine.com/2017/02/15/i-wish-id-known/

Sundance, Kyra. 2007. *101 Dog Tricks: Step by Step Activities to Engage, Challenge and Bond with Your Dog*. Beverly, MA: Quarry.

Tait, Peta. 2012. *Wild and Dangerous Performances: Animals, Emotions, Circus*. London and New York: Palgrave Macmillan.

Tait, Peta. 2016. *Fighting Nature: Travelling Menageries, Animal Acts and War Shows*. Sydney, Australia: Sydney University Press.

Tennessee State's Sportsmen's Association. 1874. "Ought Setters to Retrieve?" *Forest and Stream*, September 10, 72. Proquest Historic American Periodicals.

Times [London: England]. 1931. "Teaching Our Dogs Manners." August 10, 15. The Times Digital Archive.

Tompkins, Jane. 1985. "Sentimental Power: Uncle Tom's Cabin and the Politics of Literary History." In *Sensational Designs: The Cultural Work of American Fiction, 1790–1860*, 122–146. New York: Oxford University Press.

Tossutti, Hans. 1948. *The Companion Dog in Training: A Practical Manual on Systematic Obedience; Dog Training in Word and Picture*. New York: Orange Judd Publishing Company.

Trkman, Silvia. 2010a. *Heeling is Just Another Trick*. Slovenia: LoLaBuLand. DVD.

Trkman, Silvia. 2010b. *Tricks for a Better Bond*. Slovenia: LoLaBuLand. DVD.

Trkman, Silvia. 2011. *Ready, Steady, Go!* Slovenia: LoLaBuLand. DVD.

Trkman, Silvia. 2012. *Running Contacts That Make You Smile*. Slovenia: LoLaBuLand. DVD.

Trkman, Silvia. 2015. *Agility Diary*. Slovenia: LoLaBuLand. DVD.

Urdank, Albion M. 2006. "The Rationalisation of Rural Sport: British Sheepdog Trials, 1873–1946." *Rural History* 17, no. 1: 65–82. DOI: 10.1017/S0956793305001603

Vint, Sherryl. 2008. "Entangled Posthumanism." *Science Fiction Studies* 35, no. 2 (July), accessed March 3 2017. http://www.depauw.edu/sfs/review_essays/vint105.html

Von Stephanitz, Max. 1917. *Schooling and Training of the German Shepherd Dog*. Translated by Anne Tracy. Scranton, PA: The German Shepherd Dog Club of America.

Walsh, Julie. 2011. *Unleashed Fury: The Political Struggle for Dog-Friendly Parks*. West Lafayette, IN: Purdue University Press.

Warren, Cat. 2015. *What the Dog Knows: Scent, Science and the Amazing Ways Dogs Perceive the World*. New York, London, Toronto, New Delhi: Simon and Schuster.

Waters, Bernard. 1889. *Modern Training, Handling and Kennel Management*. Chicago: The Blakely Printing Company.

Waters, Bernard. 1901. *Training the Hunting Dog for the Field and Field Trials*. New York: Forest and Stream Publishing.

Weaver, Harlan. 2013. "'Becoming in Kind': Race, Class, Gender, and Nation in Cultures of Dog Rescue and Dogfighting." *American Quarterly* 65, no. 3 (September): 689–709. DOI: 10.1353/aq.2013.0034.

Weber, Josef. 1938. *The Dog in Training*. New York and London: Whittlesey House.

Weber, Brenda R. 2007. "Makeover as Takeover: Scenes of Affective Domination on Makeover TV." *Configurations* 15, no. 1 (Winter): 77–99. DOI: 10.1353/con.0.0020

Weber, Brenda R. 2009. *Makeover TV: Selfhood, Citizenship, and Celebrity*. Durham, NC: Duke University Press.

Weil, Kari. 2012. *Thinking Animals: Why Animal Studies Now?* New York: Columbia University Press.

Wells, Deborah L. and Peter G. Hepper. 1999. "Male and Female Dogs Respond Differently to Men and Women." *Applied Animal Behaviour Science* 61, no. 4: 341–349. DOI: 10.1016/S0168-1591(98)00202-0.

Wickham, W.A. 1891. *Practical Training of the Shepherd Dog*. Tipton, Iowa: No publisher provided. Biodiversity Heritage Library. DOI: 10.5962/bhl.title.48905.

Williams, C. H. C. 1878. *Williams' New System of Handling and Educating the Horse Together with Diseases and their Treatment*. Claremont, NH: Claremont Manufacturing Company.

Włodarczyk, Justyna. 2015. "Of Dogs and Shepherds: Sheepdog Culture and the American Pastoral." *American Studies in Scandinavia* 47, no. 1: 61–84.

Włodarczyk, Justyna. 2016a. "Canine Performance Sports in Poland: Another Look at the Dog Training Revolution." In *Free Market Dogs: The Human-Canine Bond in Post-Communist Poland*, edited by Michał P. Pręgowski and Justyna Włodarczyk, 61–85. West Lafayette, IN: Purdue University Press.

Włodarczyk, Justyna. 2016b. "Post-Communist Canine: A Feminist Approach to Women and Dogs in Canine Performance Sports in Poland." *Society & Animals* 24, no. 2: 129–152. DOI: 10.1163/15685306-12341398.

Wolfe, Cary. 2003a "Aficionados and Friend Killers: Rearticulating Race and Gender Via Species in Hemingway." In *Animal Rites: American Culture, the Discourse of Species and Posthumanist Theory*, 122–168. Chicago and London: University of Chicago Press.

Wolfe, Cary. 2003b. "In the Shadow of Wittgenstein's Lion: Language, Ethics and the Question of the Animal." In *Zoontologies: The Question of the Animal*, edited by Cary Wolfe, 1–58. Minneapolis: University of Minnesota Press.

Wolfe, Cary. 2009. "Human, All Too Human: 'Animal Studies' and the Humanities." *PMLA* 124, no. 2: 564–575.

Wolfe, Cary. 2013. *Before the Law: Humans and Other Animals in a Biopolitical Frame*. Chicago: University of Chicago Press.

Wright, Richard. 1997 [1937]. "Blueprint for Negro Writing," *The Norton Anthology of African American Literature*, edited by Henry Louis Gates Jr., Nellie Y. McKay, William A. Andrews, and Houston Baker, Jr., 1380–1388. New York: Norton.

Youatt, William. 1871 [1845]. *The Dog*. London: Longman's, Green & Co.

Young, Catherine. 2015. "'A Very Good Act for an Unimportant Place': Animals, Ambivalence and Abuse in Big-Time Vaudeville." In *Performing Animality: Animals in Performance Practices*, edited by Lourdes Orozco and Jennifer Parker-Starbuck, 77–96. London and New York: Palgrave MacMillan.

Zamora, Daniel, and Michael C. Behrent, eds. 2016. *Foucault and Neoliberalism*. Cambridge, UK and Malden, MA: Polity Press.

Žižek, Slavoj. 2001. "What Can Lenin Tell Us About Freedom Today?" *Rethinking Marxism* 13, no. 2 (Summer): 1–9. DOI: 10.1080/089356901101241640.

Žižek, Slavoj. 1997. *The Plague of Fantasies*. London & New York: Verso.

Index

Index

www.ingramcontent.com/pod-product-compliance
Lightning Source LLC
Chambersburg PA
CBHW070248260926
51849CB00001BA/46
* 9 7 8 9 0 0 4 3 8 0 2 8 8 *